ACTION AN  Y

For Gamers, Animators, and Digital Artists

Takashi Iijima

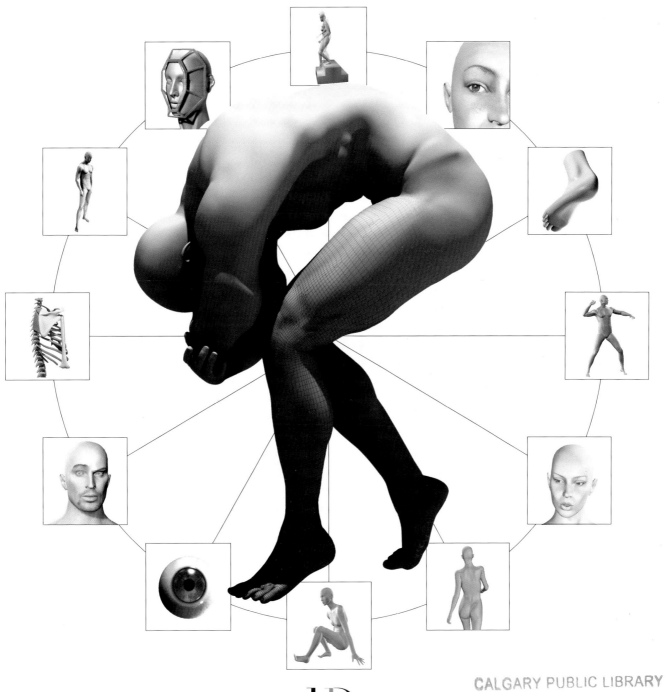

HD

HARPER
DESIGN

An Imprint of HarperCollins*Publishers*

ACTION ANATOMY

First published in 2004 by:
Works Corporation Inc.
5F Hirano Building
4-12-38 Shibaura
Tokyo 108-0023 Japan
Tel: +81 (0)3 5427-6164
Fax: +81 (0)3 5427-6596
www.wgn.co.jp

First published in English in 2005 by:
Harper Design
An Imprint of HarperCollinsPublishers
10 East 53rd Street
New York, NY 10022
Tel: (212) 207-7000
Fax: (212) 207-7654
HarperDesign@harpercollins.com
www.harpercollins.com

Distributed throughout the world excluding Japan by:
HarperCollins International
10 East 53rd Street
New York, NY 10022
Fax: (212) 207-7654

HarperCollins books may be purchased for educational, business, or sales
promotional use. For information, please write: Special Markets Department,
HarperCollins Publishers Inc., 10 East 53rd Street, New York, NY 10022.

Author: Takashi Iijima
English Translation: Chin Music Press
Typesetting and Layout: Far, inc.
Cover Design and Art Direction: Atsushi Takeda (Sourvenir Design)
Chief Editor for English edition: Rico Komanoya (ricorico)

Library of Congress Control Number: 2004109982

ISBN 0-06-073681-X

Printed in China by Everbest Printing Co., Ltd.
First Printing, 2005

About this Book

As computer graphics have become increasingly sophisticated, so have our expectations for realism. This is true for the computer games that we play, the animations that we watch, and the digital art that we enjoy. In the early 1980s, when Pac Man was at the height of its popularity, we were content to watch little creatures, whose circular bodies we only ever saw in profile, gobble each other up. Now, we want to see Onimusha's bulging muscles flex as he fiercely battles with his formidable opponents.

The trouble is, of course, that on a superficial level we are all exceedingly familiar with the human body—both its form as well as its potential for action. Whether you realize it or not, you have been studying the human body your whole life—when you brush your teeth in the morning, when you absent-mindedly see other people walking down the street, when you watch your favorite music star perform on stage. Most of us don't ever think about it, but in our everyday lives we see many different kinds of people performing every imaginable action. Consequently, when we see a CG figure who hasn't been well rendered—perhaps the hair looks funny or the motion is awkward or the lip-synch is off—we can be harsh in our critique.

This poses a real challenge for all digital creators. Not only do we need to be fluent in whatever 2D or 3D software programs we are using—whether it is Illustrator, Photoshop, LightWave 3D, 3D Studio Max, or any other of the many excellent programs that are available for digital artists—but we also need to have an extremely high level of skill in the foundations of our arts. And for many of us, this includes having a solid understanding of human anatomy.

This book will help you in your quest. It examines the human body and provides deeper knowledge of the various body parts and how they all work together, both in stop-motion as well as in action. Part 1: Structure presents the human form. This begins with complete diagrams of the human bone and musculature structures, and then continues, body part by body part, with a thorough examination of the entire body. Part 2: Action analyzes a number of typical movements. Illustrated by sequences of freeze-frame images, each of these activities is presented in great detail, with attention given to common trouble spots.

All of the chapters are connected to each other, so there's no need to read it linearly. Instead, start with any section that interests you and continue from there. We hope that this book answers all of your questions about the human body—and inspires you to create truly animated digital art.

The Editors

Contents

22 | Structure

Exploring the Construction of Human Body Parts

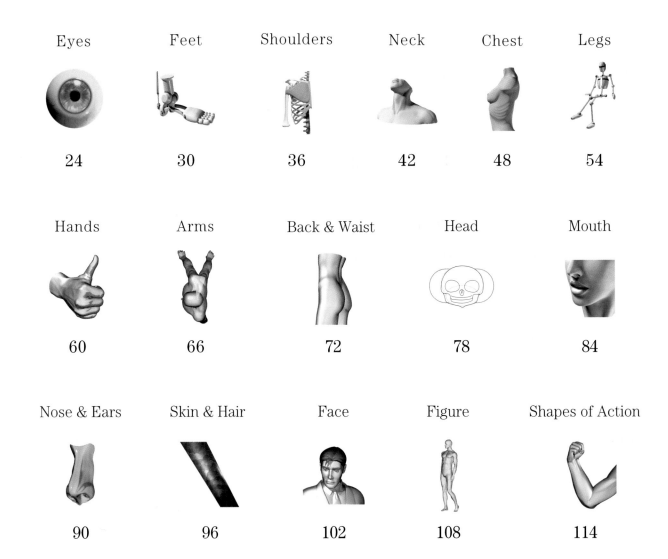

120 Action

Exploring the Mechanisms of Body Movements

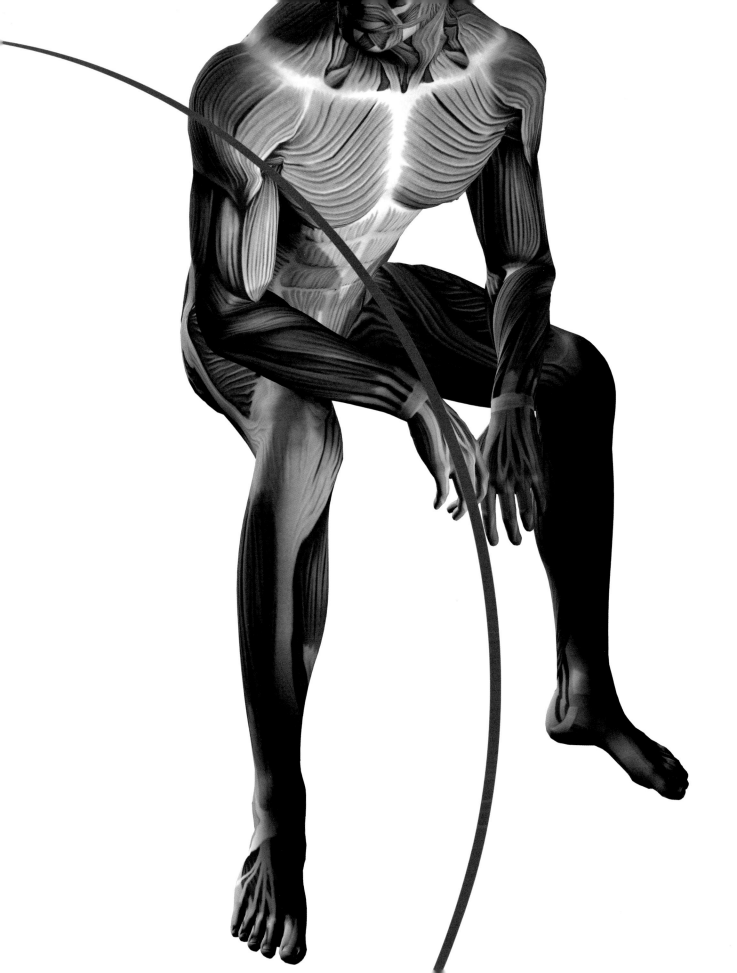

The Human Form

Bones support the core of the human body. The spine is the center, which provides flexible support. Above it is the skull, which gives central protection. The bones of the arms and legs extend in four directions. Muscles extend over the bones to cover them. Above the muscles is the flesh or skin, which shows various bulges on the surface. What form does the human body take? What mechanisms are involved in human movements? Understanding the answers to these questions is the first step in depicting the natural movements of the human body.

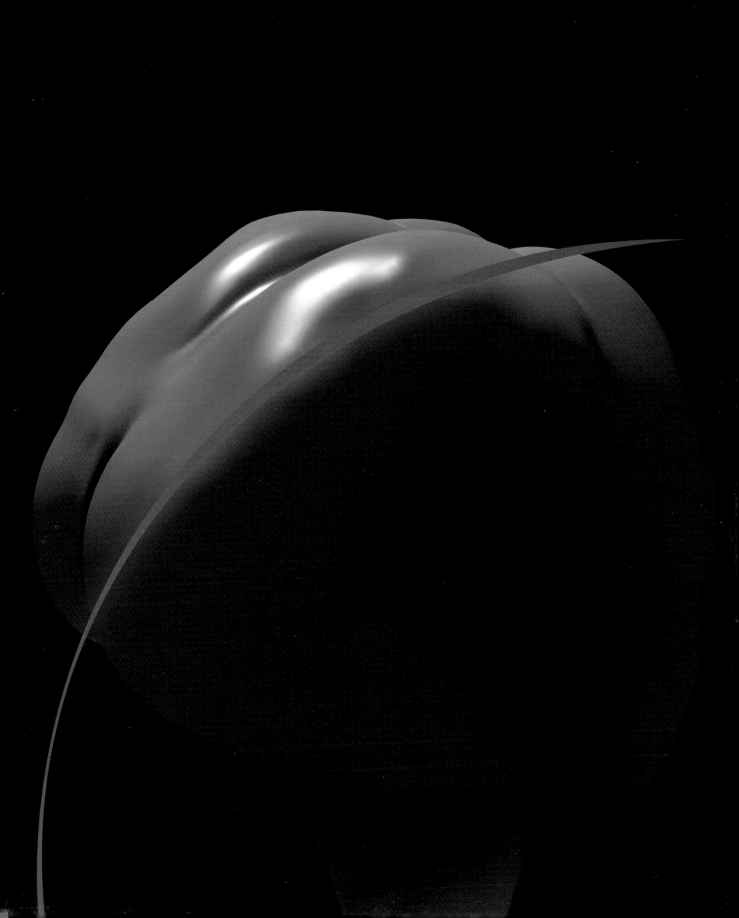

The Moving Body

Muscles provide incredible power and allow varied types of actions. When humans move, the body shape changes and the surfaces are greatly altered. Each part of the body has limited movement ability. Human bodies are heavy and change immensely when erect. After understanding the basic human structure, we need to investigate the mechanisms that help us move. Where is the center of gravity? How does someone maintain balance? When do people exert strength and where is the energy absorbed? It is important to continually and deeply investigate the basis for natural human actions.

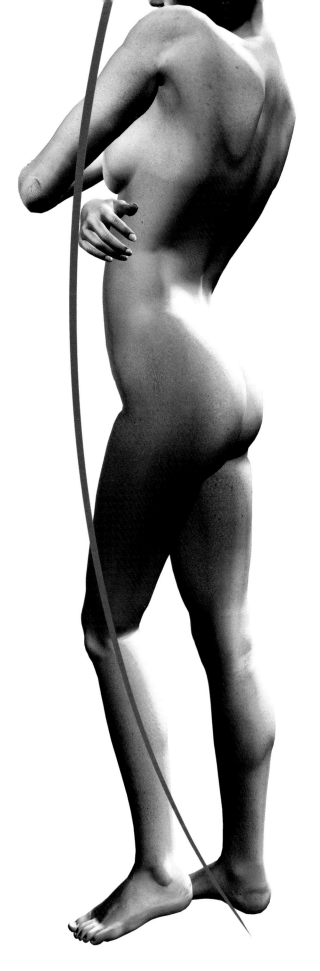

The Flow of Movement

The movements of the human body are made up of
many different currents. When a person stands, we
can see the center of gravity pass from the head to the
ground. A smooth relationship links the series of
actions. Most actions don't end in one movement;
they're connected to a succession of actions and
various body parts. When the actions are woven, fluid
movements emerge. There are two types of
movements: those that are intentional and those that
comply with physical laws. An example of the former
is when a person extends his or her arm to open a
door. An example of the latter is when the person's
arm releases the doorknob and drops after the door is
opened. To understand the complex movements of
the human body, one must probe beyond the currents
and focus on the intention of the person.

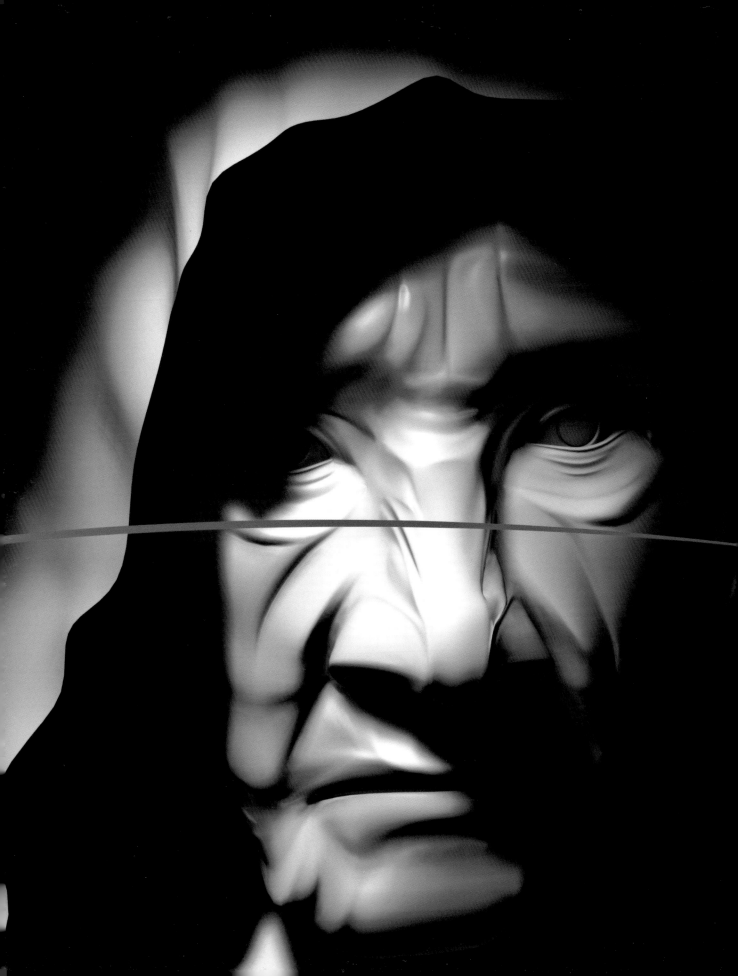

Awareness and Habits

People have idiosyncrasies. The factors related to a person's actions include age, gender, occupation, environment, character, and race. Even if you can't see a person's face, you can identify him by his gestures. A peculiar habit can make a very strong individualistic impression. An old person's face reveals the experience of time, which has been carved into it. Also, faces express something that is seared into one's memory, and which can't easily be summed up in words. Each person has a different posture. One trivial gesture is connected to many factors, such as taste, aesthetics, situation, knowledge, feeling, and condition. Actions become alive when these factors are interwoven. A model that may look real but doesn't move is nothing more than a wax doll. Movement helps us to feel life. One simple movement can be overwhelming, as it brings us closer to the mystery of life.

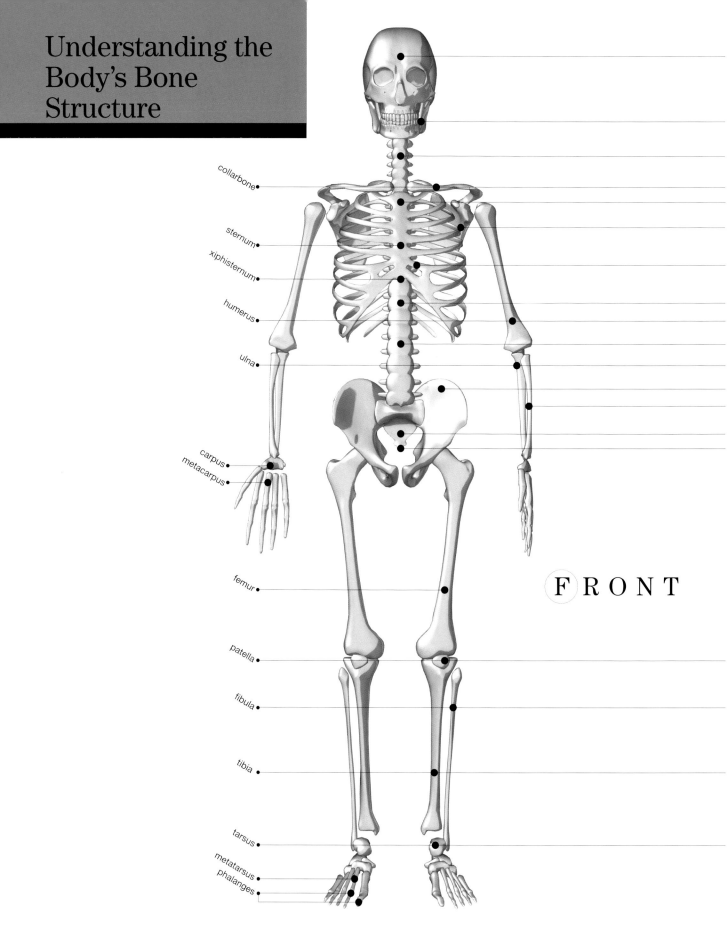

collarbone

sternum

xiphisternum

humerus

ulna

carpus
metacarpus

femur

patella

fibula

tibia

tarsus

metatarsus
phalanges

FRONT

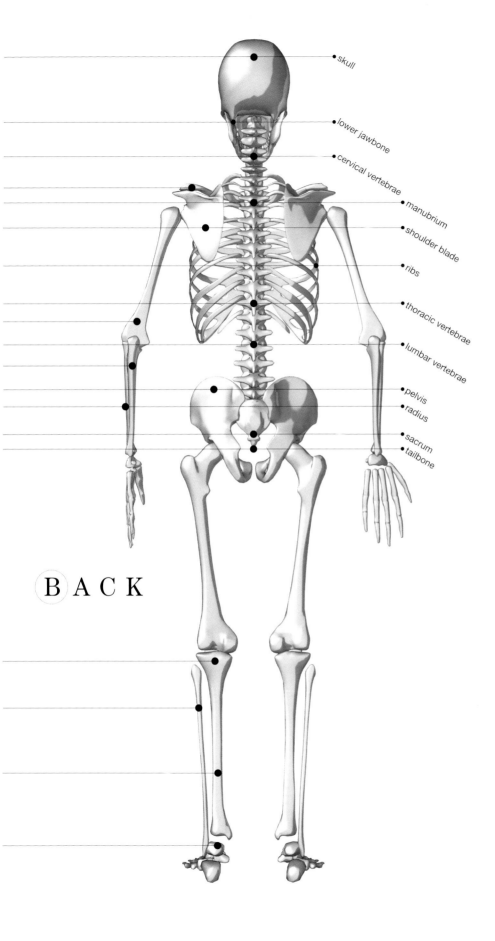

skull

lower jawbone

cervical vertebrae

manubrium

shoulder blade

ribs

thoracic vertebrae

lumbar vertebrae

pelvis

radius

sacrum

tailbone

BACK

The human body is comprised of roughly 200 bones, all of which form a strong column to support the body and provide an armor to help it move. Some bones can actually be seen right below the surface of the skin; they often look different when in motion. Bones manifest themselves differently depending on whether they are beneath soft layers of fat or muscle, or are protruding. The skin gets taut where the muscle is working, creating many forms of expression. A bone's form can change greatly during movement. Bones can be divided roughly into two parts: hard bones and soft cartilage. Cartilage refers to areas between the joints, at the base of the frontal ribs, or on the nose, for example.

15

skull

Pages 78-83

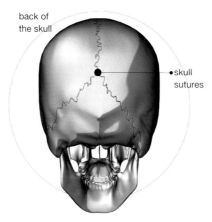

back of
the skull

skull
sutures

We begin with the skull. Much like a jigsaw puzzle, the skull is a fusion of many small bone parts. A baby's skull is not yet fully fused. You can actually feel the gap with your fingers, although this is not advisable. The area where the greatest action takes place is the jawbone. The jawbone opens up by sliding downward diagonally. This movement is easy to confirm because it creates a small bulge at the base of the jaw by the ears. The head is also the one place where certain bones are visible—in the form of 16 teeth each in the upper and lower jaw.

chest

Pages 48-53

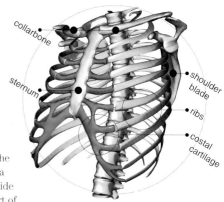

collarbone

sternum

shoulder
blade

ribs

costal
cartilage

The flat bones of the shoulder blades have a unique shape. They slide freely with the support of various muscle groups. The shoulder blades are sometimes visible, depending on the motion of the arms. The collarbones, which help the shoulder blades move, curve dramatically backward. Collarbones are small and fragile, but they are indispensable tools of expression because of their prominence above the chest area. The breastbone is divided into the manubrium, the sternum, and the xiphisternum. There are 12 ribs on either side. The costal cartilage extends from the thoracic vertebrae to the sternum. The ribs, due to the elasticity of the costal cartilage, can absorb external impact and expand upon breathing. Rib bones are thin and form a cage around the lungs and heart.

back

Pages 72-77

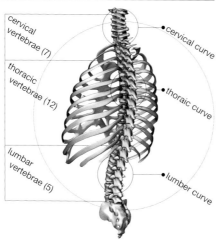

cervical
vertebrae (7)

cervical curve

thoracic
vertebrae (12)

thoraic curve

lumbar
vertebrae (5)

lumber curve

The backbone extends from the base of the skull to the hip bone. Some 24 small bones are connected like cornices. The first seven vertebrae are called the cervical vertebrae and curve forward. The next 12 are called the thoracic vertebrae. They curve backward and form the base of the ribs. The remaining five are called the lumbar vertebrae, which curve forward. From the cervical vertebrae, the bones become gradually larger until they reach the lumbar vertebrae, where they become slightly smaller. Radiating from the backbone are several protuberances: one on each side, one facing down, one going in the opposite direction (the upper joint projection), and yet another one below that (the lower joint projection). These become visible when you round your back. The curves of the spine serve as springs for the entire back, designed to support the head and upper body.

waist/hips

Pages 72-77

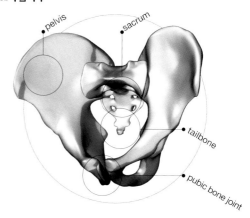

pelvis

sacrum

tailbone

pubic bone joint

The pelvis bone extends to the left and right, anchored in the middle by the sacrum. At the bottom, the two sides meet, sandwiching the cartilage. The sacrum is part of the backbone. Its many surface indentations are reminiscent of a reptile. At the lower tip is the tailbone—a remnant of the tail that evolution rendered a vestige. The pelvis opens at the top, its shape different according to gender. Look at the male pelvis from an opening at the lower part, and you'll see that it is triangular and narrowly tilted to the front. A woman's pelvis is rounder to aid in childbirth. The human hip bone is larger and better developed than that of four-footed animals because it has to support the upper body.

arms, hands

Pages 60-71

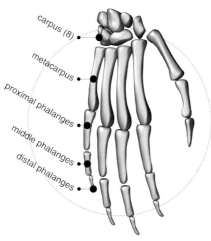

The humerus is in the upper arm, and the ulna and radius are in the forearm. At the base of the hands are a group of eight bones called the carpus. Starting from the thumb side, they are the trapezium, trapezoid, scaphoid, capitate, lunate, hamate, triquetrum, and pisiform. The palm of the hands has five bones, the metacarpus. Each finger is comprised of three bones. The thumb has only two bones. Fingernails have a dual function. They protect the fingers while expanding their functions. Nails facilitate detailed finger work that soft fingertips alone are unable to perform. Thanks to the nails, we can pick up small objects. Their function is similar to bones, but nails are simply hardened skin.

legs, feet

Pages 30-35, 54-59

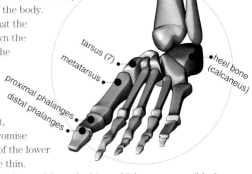

The bones of the feet constantly bear the weight of the body. Gravity dictates that the farther you go down the human anatomy, the stronger the bones and the better developed the muscles should be. And yet, so as not to compromise agility, the bones of the lower part of the legs are thin. Energy flow can be observed from the hips, which are responsible for maintaining the body's equilibrium. Energy travels from outside of the hips toward the back through the inside of the knees, ending at the Achilles tendon. This spiral flow of energy is significant in understanding the foot's structure. The feet consist of seven tarsal bones—the calcaneus, talus, navicular, cuboid, lateral cuneiform, intermediate cuneiform, and medial cuneiform; five metatarsal bones; and three bones on each of four toes, with just two on the big toe. The bones of the feet are considerably larger and stronger than those of the hands. The bones of the big toe are especially large.

How Joints Are Constructed

The basic joints are displayed in the drawings below. Roughly speaking, most joints move in the way depicted by the drawings on the upper left. Flex, and the muscle will contract; relax, and it will return to normal. That's all it can do. To get a joint to bend and return to position—or to get it to maintain any angle—it has to be pulled from both ends. The movable parts of the bones form a pair, and the ligaments, which are stretched in all directions, provide assistance. As the drawings at the bottom of the page show, there are six basic joint shapes: spherical, oval, axle, hinge, saddle, and plane. The spherical joints move on multiple axes and are found mainly in parts of the body that make big movements, such as the shoulders and thighs. Oval joints look like spherical joints, but they move along a smaller axis. They can be found in the wrists, for example. Axle joints allow for twisting and can be found below the elbows and knees. Hinge joints are found in elbows, knees, and other similar places. Saddle joints have slightly odd forms and can be found in the palm of the hand at the base of the metacarpus, for example. Plane joints are found along the backbone, collarbone, and other such places. Many joints have cartilage between the bones to protect against chafing.

Basic Model for Joints

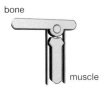

Joint Shapes

spherical joint

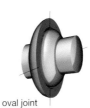

oval joint

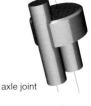

axle joint

Parts That Support a Joint's Movement

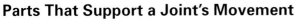

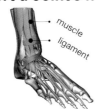

hinge joint

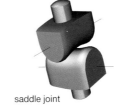

saddle joint

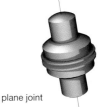

plane joint

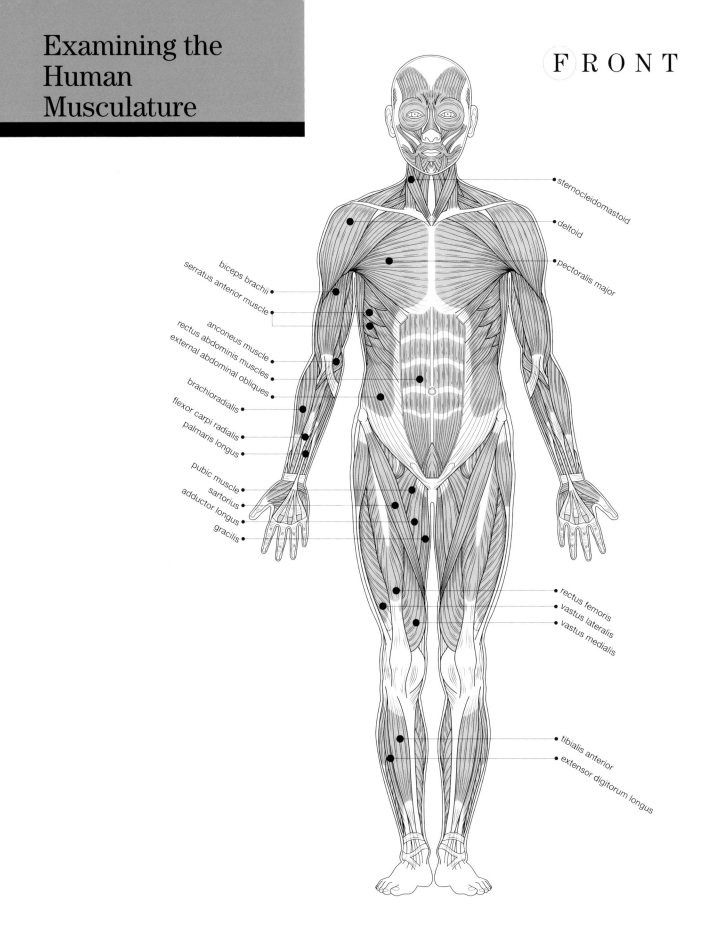

sternocleidomastoid

deltoid

pectoralis major

biceps brachii
serratus anterior muscle

anconeus muscle
rectus abdominis muscles
external abdominal obliques

brachioradialis

flexor carpi radialis
palmaris longus

pubic muscle
sartorius
adductor longus
gracilis

rectus femoris
vastus lateralis
vastus medialis

tibialis anterior
extensor digitorum longus

B A C K

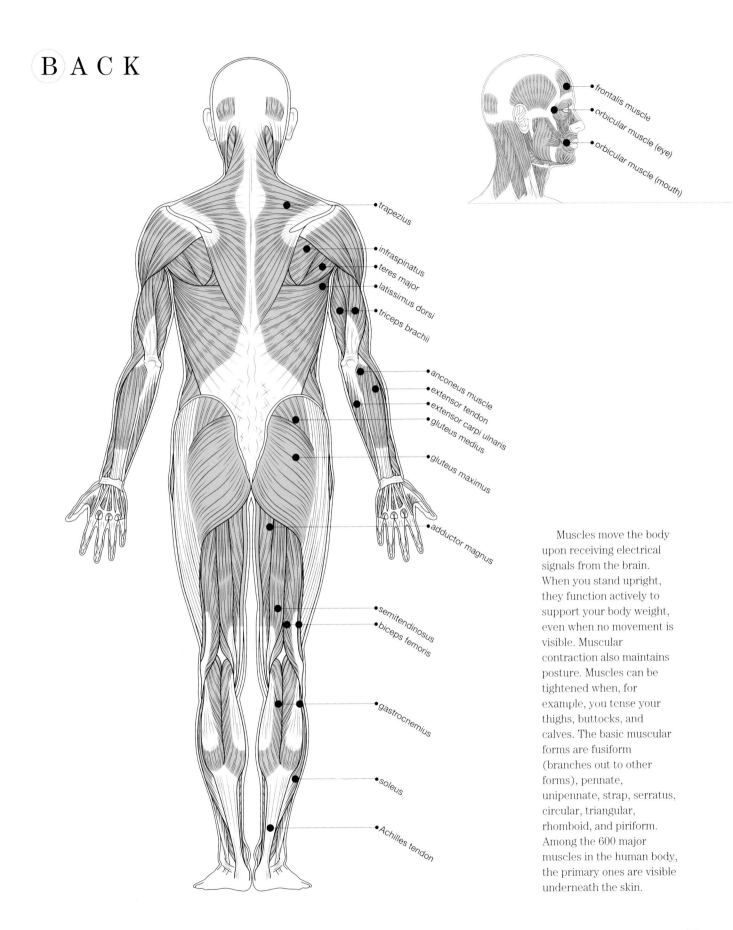

- trapezius
- infraspinatus
- teres major
- latissimus dorsi
- triceps brachii
- anconeus muscle
- extensor tendon
- extensor carpi ulnaris
- gluteus medius
- gluteus maximus
- adductor magnus
- semitendinosus
- biceps femoris
- gastrocnemius
- soleus
- Achilles tendon
- frontalis muscle
- orbicular muscle (eye)
- orbicular muscle (mouth)

Muscles move the body upon receiving electrical signals from the brain. When you stand upright, they function actively to support your body weight, even when no movement is visible. Muscular contraction also maintains posture. Muscles can be tightened when, for example, you tense your thighs, buttocks, and calves. The basic muscular forms are fusiform (branches out to other forms), pennate, unipennate, strap, serratus, circular, triangular, rhomboid, and piriform. Among the 600 major muscles in the human body, the primary ones are visible underneath the skin.

head, neck

Pages 42-47, 78-83

There are basically two sorts of muscles in the head: weaker ones that change the facial expression and more powerful ones that give the jaw its great strength. The muscles that move the jaw are the temporalis, which spread across the sides of the head, and the masseter, which connects the cheeks and the mandible, or lower jawbone. They work together to both unleash power and create delicate movements. The muscles that change facial expressions are around the eyes, mouth, and brow. The eye muscles, which need to react quickly, are strong and thick. The sternocleidomastoid extends from behind the ears to the middle of the collarbone. The big muscle at the back of the neck is the trapezius; it helps support the neck and shoulders. It is said that stiff shoulders are caused by poor circulation, and muscle pain is brought about when the fibers that make up the muscle swell in the process of thickening.

arms, hands

Pages 60-71

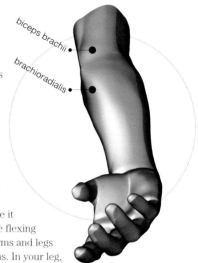

The biceps brachii is one of the more prominent muscles—it is the muscle we flex when we want to show those well-developed biceps. It allows us to bring our arms toward us. From the elbow to the wrist, much of the muscle is for moving the fingers. Relax your arm, then make a fist. Squeeze it tight. You should feel the flexing throughout your arm. Arms and legs have similar compositions. In your leg, the thigh is where a well-developed muscle can stand out like the biceps. The shin contains a complex array of muscles like the lower arm. In the hand, there are a lot of muscles on the palm. The muscles you can see on the surface are entirely related to the thumb's movement. Even the little abductor digiti minimi muscles play important roles in shaping the hand.

shoulders, chest, abdomen, back

Pages 36-41, 48-53, 72-77

Front

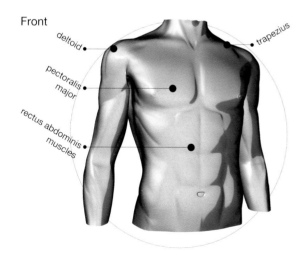

Back

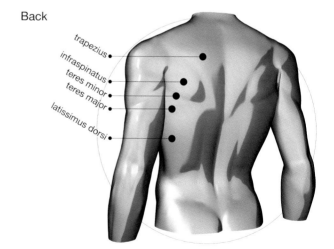

A man's chest muscles spread from the center of the ribs, where the breastbone is, to each shoulder. Most of these muscles help the arms move. The diaphragm, the internal intercostals along the collarbone, and the transversus muscles in the abdomen are not visible. In the back, the teres major, infraspinatus, and teres minor create protuberances. For the back to move flexibly, a lot of muscles must function in unison. The sentence "He has no backbone" conjures up someone who is unwilling to take action. If the backbone were really taken out, a human would hardly be able to move at all. The upper part of the backbone reaches to the neck, making it a vital bone. It acts as a spring to protect the skull and brain. Finally, the rectus abdominis muscles, planklike muscles in the abdomen, protect the internal organs and create blocklike protuberances.

hips, legs, feet

Pages 30-35, 54-59, 72-77

Hip and leg muscles go together like shoulder and arm muscles. The pectoralis major in the chest is like the gluteus maximus in the buttocks. In the thighs, most of the muscles exist in a complicated tangle. Notice the change in the body's surface made by muscles such as the sartorius, which spreads from the outer hips to the inner knees, the biceps femoris on the back of the thighs, the semitendinosus, and the semimembranosus. On the calves, a muscle called the gastrocnemius forms two mounds. Below that is the soleus, which connects to the Achilles tendon. It is said that even a warrior cries when hit in the shins, a place with little muscle or fat and with the bone just below the skin's surface. But muscles to move the leg and the toes are found here, so the surface does change a little. Most of the muscle on the foot is on the instep. The toes play a vital role in balancing. For example, muscles in the calves allow people to straighten their toes and keep from falling.

Front

sartorius
rectus femoris
vastus lateralis/ medialis
tibialis anterior
soleus

Back

gluteus maximus
biceps femoris
gastrocnemius
soleus
Achilles tendon

Describing and Explaining Movement

Yielding to Gravity

Most muscles run counter to the joints they support. That means stretching out the joints requires muscle work. The exception is when gravity is at work. The force of gravity depends upon the weight and mass of the matter in question. It takes tremendous strength to defy gravity. On the other hand, you could let your body give in to it by rolling around on the floor or letting your arms drop. When you let your arms hang, your arm muscles will relax. Then the weight of your arms will shift to your shoulders. You can curl your back to shift that weight off your shoulders, but that again only puts the onus on some other part of the body. When we are in the same position for hours, we seek ways to shift the weight of our arms—by putting our hands in our pockets, for example. When you cross and uncross your legs or rest your elbows on the table, pay attention not only to the muscles that flex and relax, but also to the changes incurred by flattened muscles and body fat.

The body supports the weight of the arms.

Motionless Tension

The Thinker is a famous work of art by the 19th-century sculptor Rodin. The subject is in deep thought. Even if you don't know the title of the sculpture, you can tell what the subject is doing just by looking at him. The way he rests his elbow on his knees, the way his right shoulder curls in, and the tension seen in the shoulder muscle that supports the dangled left arm all precisely depict the muscles at work. His brow is furrowed. His entire body, including his face, is gripped with muscular tension. We often cross our arms when in thought. It helps to disperse the weight of the arms and allows us to relax. It is also said that crossing the arms creates just the right amount of tension in the arms, and that stimulates the brain. The simple act of thinking, which on the surface may seem unrelated to movement, creates muscle tension that signals the thinking process.

Momentary Ripples

When you run or jump, the moment you land, your leg muscles—starting with the calves, then the thighs—tighten momentarily. You can tell that the legs have landed by the way the muscles try to control the impact, creating waves, like ripples in the water. The largest muscle movement is seen upon landing; the effect weakens after that.

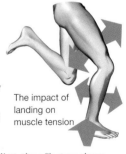

The impact of landing on muscle tension

The Power in Repose

Michelangelo's statue of David captures the young man gathering strength moments before his battle with Goliath. David's posture suggests that he may be in a state of calm repose in the way he rests his weight on one leg. And yet, his entire body emanates the energy that foretells his next course of action. Intentionally sculpting the head, feet, and hands proportionately larger than the rest of the body is not just a trick of the trade to express physical power. The precise muscular portrayal, especially the contrast between tense and relaxed muscles, is also an excellent depiction of the flow of energy. Attention to the details of muscles at their height of power can even demonstrate inner resolve.

21

Part 1

Structure

Exploring the Construction of Human Body Parts

The goal of this book is to portray people as naturally as possible in animation. To do this, we first have to understand the basic structure of the human body. The body's structure is complex, and many diverse parts have to cooperate to create movement. The body is also efficiently constructed so that each part can maximize its role and function. If we take the time to investigate, we will find that even the parts we take for granted contain some surprising revelations.

A word of warning

The length and proportion of the various body parts are based on an actual 25-year-old woman and 28-year-old man. These two people are not especially representative of average body types, so please consider the drawings as samples and nothing more.

eyes

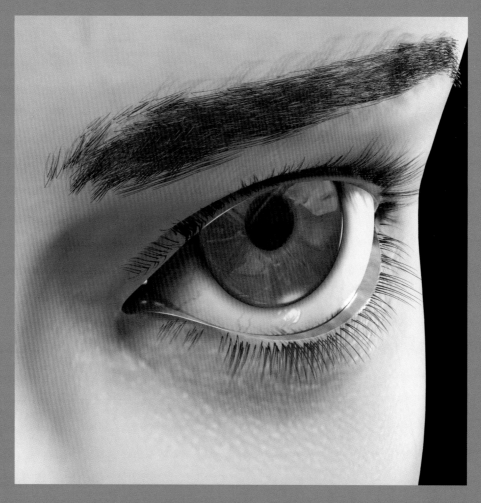

The eye is a very mysterious organ. Its function is for seeing, which is achieved by gathering reflected light, projecting that onto the retina, and transmitting that image to the brain. That is really all it does, so why is it often referred to as the "window of the soul"? By observing what another person is looking at, you may be able to tell something about what interests that person. How a person looks at an object helps you understand his or her relationship to that thing. The eyes are capable of showing a wide range of emotions.

Related pages   p. 90 nose & ears   p. 102 face   p. 170 looking back

e
eyes

Structure

Exploring the
Construction of
Human Body Parts

Part 1

eyes

eye proportions ☞1

Example: 25-year-old female

The eyeball is not a perfect sphere. The area surrounding the pupil forms a small mound. The eyeball is about 1 inch in diameter, with the pupil approximately half of that.

iris

crystalline lens

approximately .4 inch

cornea

approximately 1 inch

pupils ☞2

The pupil is slightly larger than .4 inch in diameter. It is located in the center of the eye. The iris controls the amount of light that enters the eye by adjusting the diameter of the pupil.

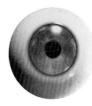
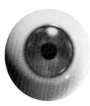
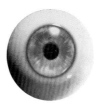
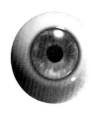

eye muscles

The eye has six muscles. They control eye movement mainly by pulling the eyeball from behind, allowing for quick and precise movements.

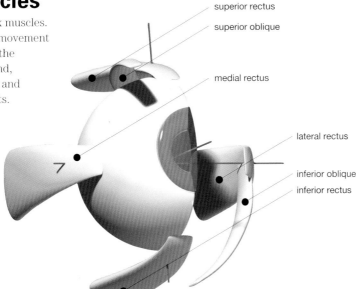

superior rectus

superior oblique

medial rectus

lateral rectus

inferior oblique

inferior rectus

☞1. **The size and textural qualities of the eye**

Age, rather than individual differences, is a factor in the size of the eyeballs. The inside of the eyeball is filled with clear watery gel that keeps the eye constantly moist and soft.

☞2. **The shape of the pupil**

There are few individual differences in the pupil's size. The eyes appear larger or smaller depending on the size of the opening. The surface of the eye is smooth and curved, but the surface of the iris has delicate striations that rise gradually in rough patterns toward the center. The border between the white of the eye and the colored part of the eye is blurry. Normally, the pupil appears black. But when light enters the pupil, the retina will appear red, as it often does with a camera flash.

☞2. **The color of the iris**

Melanin determines the color of the iris. The greater the amount of melanin, the darker the iris, as in black or brown eyes. Less melanin would result in gray or blue eyes. Lighter-colored eyes do not shut out light as effectively as darker ones, so the same amount of light would appear brighter to lighter-colored eyes. That's why people with light-colored eyes see better in dim surroundings and why we often see people with blue eyes squint and shield their eyes with their hands when they step out into strong sunlight.

eye whites

The surface of the white part of the eye is a bit bumpy and not as smooth as the pupil. Many capillaries run across the surface. Blood pools to these capillaries when the eyes become tired or lack sleep, making the eyes red. Eye whites are basically white with a slightly bluish tint, but can change according to one's health. The constant secretion of tears creates a small pool of water between the eyes and the eyelids to keep the eyes moist.

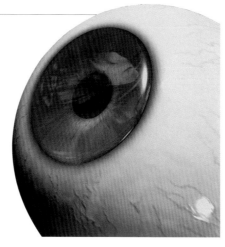

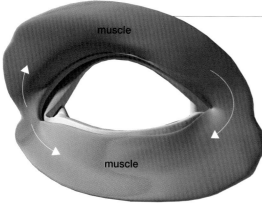

muscle

muscle

eyelids

A thin layer of muscle circles the eyes beneath the eyelids. The lower lid moves in conjunction with the upper lid. The lower lid has very little fat, so something as simple as pulling an all-nighter can burn that fat, revealing the veins beneath the skin's surface and creating dark circles. The shape of the eyelids has subtle individual differences.

animal eyes ←3

The irises in rabbits' eyes lack any pigmentation. Because of their transparency, the retinas, which are red, are visible to the naked eye. The cat's pupil is vertically elongated, like a slit. This allows for rapid light adjustment. Cats that are nocturnal animals need eyes that help them navigate nimbly through the dark of the night. The pupils are vertical to match their line of vision—that is, up and down—enabling cats to climb trees and fences. Horses, on the other hand, have pupils that are stretched horizontally to adapt to their life on flat lands, which inclines them toward horizontal vision.

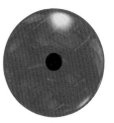

rabbit

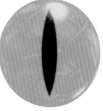

cat

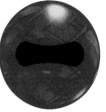

horse

eyelashes

Eyelashes are about .4 inch long. The upper lashes tend to be thicker and clumped together, numbering about 100 to 150. There are about 50 to 75 lower lashes. Eyelashes appear to grow parallel to each other, but they actually overlap to protect the eyes from dust. The number of eyelashes varies greatly from person to person. The more eyelashes one has, the more noticeable they are.

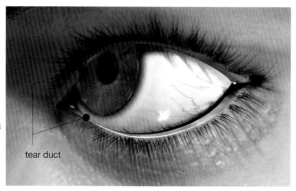

tear duct

←3. Animal eyes

Some animals, such as birds, have eyes that don't move. Birds rotate their heads to see. From the time that a bird moves its head to see to when it stops to focus, the subject becomes blurry. To cut down on this time, the bird's movements tend to be very linear. The way human eyes move is similar to the way a bird's head moves. The only time the movement is smooth is when the eyes are following the motions of their subject.

e
eyes

Structure

Exploring the
Construction of
Human Body Parts

Part 1

eyes

eyebrows ☛4

One eyebrow has roughly 650 hairs that measure approximately .4 inch.

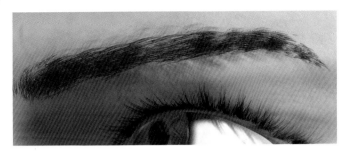

iris movements ☛5

The iris changes the size of the pupil in the presence of light. Under strong light, the diameter shrinks to about .1 inch. In total darkness, the pupil dilates to about .3 inch.

.3 inch

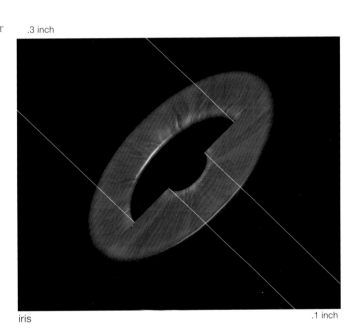

iris

.1 inch

☛4. Eyebrows

It is harder to read the expression on the face of someone without eyebrows. This inability leads to insecurity. Generally, the Japanese have flatter features than people of European descent, so their eyebrows are located higher up on the face.

☛5. Iris movements

The iris is activated by changes in the amount of light that enters the eyes and when the beholder is looking at something intently. To break it down even more, there are three categories: direct light, indirect light, and close-range reflex.

Direct light reaction

This is a swift reflex. When a very bright light enters the eyes, the pupil contracts within a second, then relaxes a little to adjust.

Indirect light reaction

This refers to when light enters one

eye and the pupil of the other eye also contracts in reaction.

Close-range reflex

When looking at something at close range, the pupil will contract to focus.

eyelid movements ☞6

Blinking and eyeball movements go hand in hand; blinking keeps the image crisp and clear even while the eyes shift their focus. The left set of pictures shows the eye blinking while focusing straight ahead. The right set shows the eye blinking as it shifts its focus to the left.

closing

position

0.66 seconds/2 frames

opening

position

0.5 seconds/15 frames

slow return to full position

eyeball movements ☞7

The left and right eyes focus together on an image. That means the eyes will become cross-eyed when focusing on an image in close range.

Also, the right eye and the left eye do not move in symmetry. They only follow what they see, creating differences in how they move.

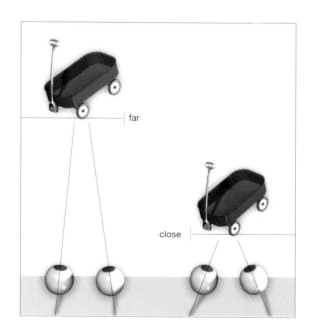

far

close

☞6. Eyelid movements

Blinking can be divided into two parts: cyclical (an average of 20 times per minute for an adult male; 15 times for the adult female) and reflexive, to protect the eyes from external matter. Winking is a deliberate form of blinking. See the pictures above. If we measured one second in 30 picture frames, blinking takes up about 15 of those frames. The eye closes in four frames, opens in 11 frames. Furthermore, the eye closes rapidly in three frames, and shuts tight in one frame. Then the eye opens in six frames and returns slowly to its original position in five frames. The eyeball lags slightly behind the eyelid in turning down, but returns to its original position quicker than the eyelid. The lower eyelid is also moving during this time. In the first four frames, it rises a little; it goes back down in the next six frames; and then it stops.

☞7. Eyeball movements

Eyes move instantaneously. When they move from left to right, blinking keeps the brain from registering the blurry vision that results from the sweeping motion. Blinking brings the gaze down and to the center. Then the eyes look to the side while the eyelids reopen. One quick, dark moment takes out the blur. Eyelids move together, too. If you look up, your upper eyelids will move up, and if you look down, both your lower and your upper lids will move down. Lifting your upper eyelid while looking down is virtually impossible. It feels like you are lifting your eyelids, but you are actually only lifting your eyebrows.

e
eyes

Structure

Exploring the
Construction of
Human Body Parts

Part 1

eyes

ranges of eyeball movements

Looking straight ahead, the range of movement for the eyes in (a) is 20 degrees above and 30 degrees below that point, totaling 50 degrees. In (b), the range becomes 30 degrees to the right and 50 degrees to the left, totaling 80 degrees. The natural position of the eyes—when the eyes are not focusing on anything—is not along a horizontal or vertical line. Rather, the eyes rest at a somewhat lower gaze, looking slightly outward.

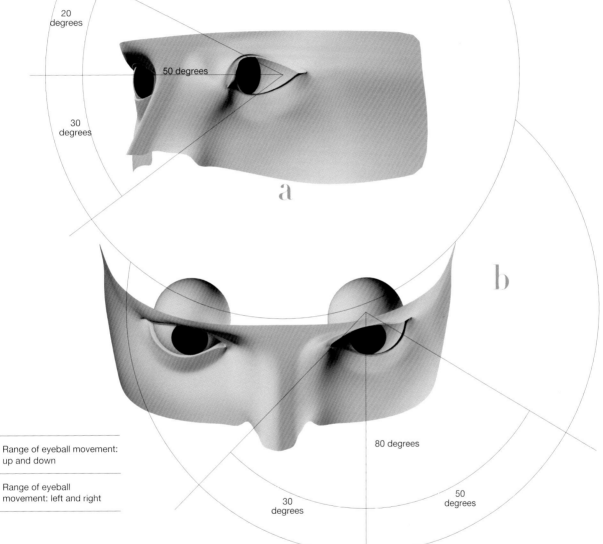

a Range of eyeball movement: up and down

b Range of eyeball movement: left and right

Habits of the eyes

We rub our eyes when we get tired. Our brain sends a message that it wants to rest and decreases the secretion of tears, drying out the eyes. In a relaxed state when the eyes are not focusing, they move slightly down and toward the center, and the pupils dilate a little. It is the most comfortable position for the eyes. In conversation, we look at the person talking. In doing so, we send the message that we are listening. We habitually protect our eyes. They are symbols of our awareness, and they continue to be mysterious organs.

feet

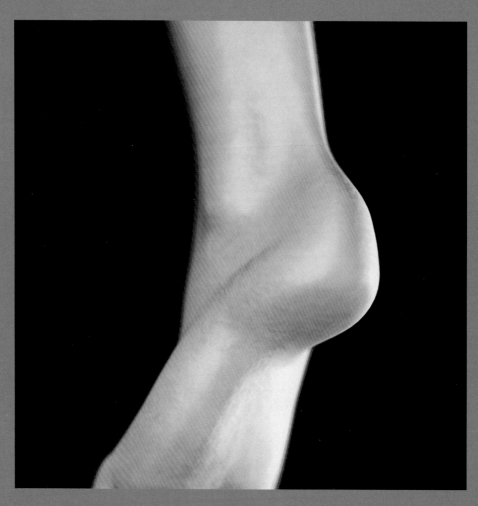

On Earth, humans are pulled by 1g of gravity. The bottoms of the feet sustain the body's weight, and the strength required to walk or run rises exponentially. The legs are said to be three times stronger than the arms, but the different parts of the feet are much stronger than the hands. Bend your big toe and see if you can straighten it using only your arm strength. It's difficult with both arms, isn't it? The other toes also have great strength. We can see why the foot is invested with such power when we think about the role it plays and the way it moves. This chapter will explore the foot, from the anklebone down.

Related pages   p. 54 legs   p.108 figure   p.114 shapes of action   p.122 getting up   p.134 walking
p.140 climbing up and down   p.146 running   p.152 jumping   p.164 standing

f feet

Structure

Exploring the
Construction of
Human Body Parts

Part 1

feet

foot bones ✏1

The bones of the foot can be divided into three parts: the tarsus, where the heel is; the metatarsus, on the instep; and the phalanges, where the toes are.

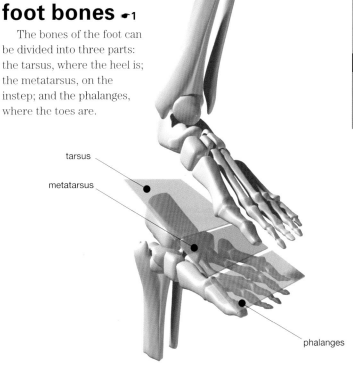

tarsus

metatarsus

phalanges

foot proportions

Example: 25-year-old female
From heel to toe, the foot is about 10 inches long. The bottom of the heel to the inside anklebone measures about 3.2 inches. It's about 2.8 inches to the outside anklebone.

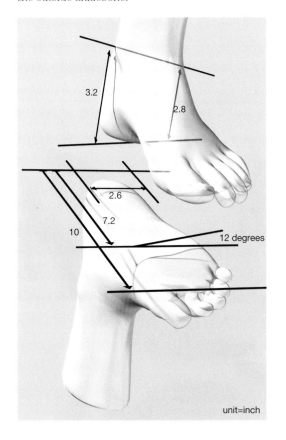

3.2

2.8

2.6

7.2

10

12 degrees

unit=inch

foot structure ✏2

The complicated muscle structure allows the foot to move freely. The anklebone slants downward, favoring movements to the inside.

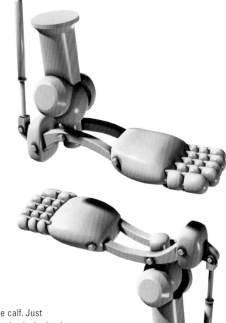

✏**1. Supporting humans: the foot and its bones**

The tarsus in the heel is made up of seven bones. The biggest ones are the talus and heel bone. There are also five blocklike bones. The metatarsus on the instep consists of five bones, and the phalanges are made up of 14. The big toe has two bones; the other toes have three

each. The big toe, like the thumb, has one less joint than the other digits. It is stronger and can move more freely than the other toes.

✏**2. The muscles and tendons that create agility**

The foot's muscles are a complex group. The one that stands out the most is the soleus, extending from the

Achilles tendon to the calf. Just below the Achilles tendon is the heel bone; the anklebone rotates the foot up and down. There are also important muscles in the bottom of the foot, the ankle, and the toes.

The muscles around the toe joints are toward the bottom. The complicated mesh of muscles in the ankle allows it to move freely.

foot surfaces

The instep doesn't come into much contact with air or sunlight, so the skin is lighter and soft. With the exception of the arch, the bottom of the foot has thick skin from pressure and chafing.

shoes ☞3

The standard pair of shoes is reinforced along the fulcrum. The protection is made even stronger on the heels, which carry the most weight.

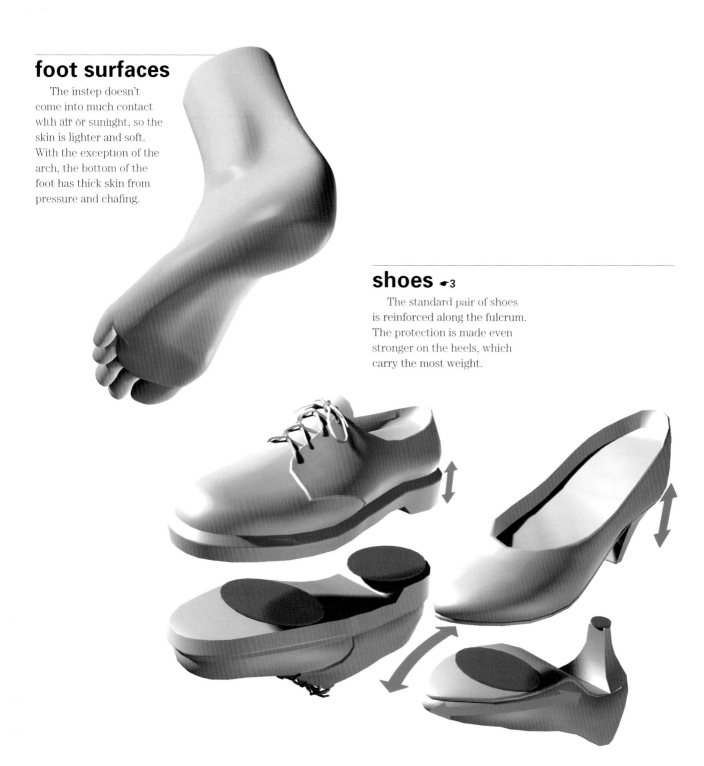

☞3. The intimate relation between the shape of the foot and the shoe

An intricate web of thick blood vessels is visible on the surface of the foot; some of the vessels wrap around to the bottom. The toenails are much smaller than fingernails. They become brittle as they grow out and grow downward on the sides. But toenails vary widely depending on the type of shoes a person wears, the way he or she walks, and so on. And it's not just the nails that reflect individual differences in the feet: there are rigid big toes, hammertoes, calluses, and other manifestations.

Today, people spend most of their time wearing shoes. Whether the shoes fit the feet or not, they are like part of the body, and they can greatly affect the shape and function of the feet. For the leg to smoothly move forward, the opening of the shoe has to widen a lot. When women wear high heels, which raise the heel in order to tighten the calf, they are on their tiptoes. But in the quest to make their feet seem slimmer, women can wear shoes that produce rigid big toes and calluses on the smaller toes.

f feet

Structure

Exploring the
Construction of
Human Body Parts

Part 1

feet

toes ☛4

When the two arches on the foot can't bear the load, the five toes support the body as it moves forward. The weak-looking little toes are actually little so that they can exert great strength at their tips. When the tips of the toes can't bare the load, the other foot reflexively takes a step forward.

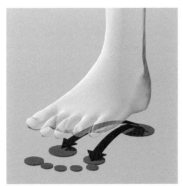

regaining balance ☛5

The toes, the bottom of the foot, and the heel each help a person balance as he uses his center of gravity and his reactions to balance.

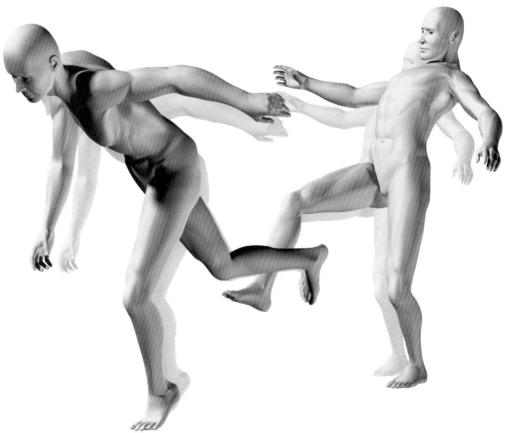

☛4. The bottom of the foot

When a person is standing still, the foot is structured so that three parts of the longitudinal arch are in contact with the ground, and balance is flexibly maintained. The foot's strength lies in the structure of the arch that helps disperse pressure from the top. Because a foot has two arches, it is able to withstand heavy pressure from many angles.

☛5. Maintaining and reacting to one's center of gravity

When you reflexively take too big of a step forward, shifting your center of gravity, you tend to overcompensate by extending diagonally. At this time, all of the toes are working to support your body. When shifting sideways, people tend to stand with their feet apart, so there is little likelihood that they will need to further stretch their

legs. Normally, to keep from falling when stepping forward, you stick out your hips, force your head up, turn toward the inside, and raise one foot into the air behind you.

When you are trying to maintain your center of gravity while falling backward, you quickly step back with one foot. The heel bears a lot of weight even when you're just standing, and it is natural for it to

compensate like this. But your foot doesn't just step back. To keep from falling, you clench your jaw, thrust your hips forward, extend your arms to the side, and lift your other foot in the air.

keeping balance ☞6

Walking is more than just an activity where the legs propel the body. It's more like continuously and skillfully balancing ourselves as we fall forward.

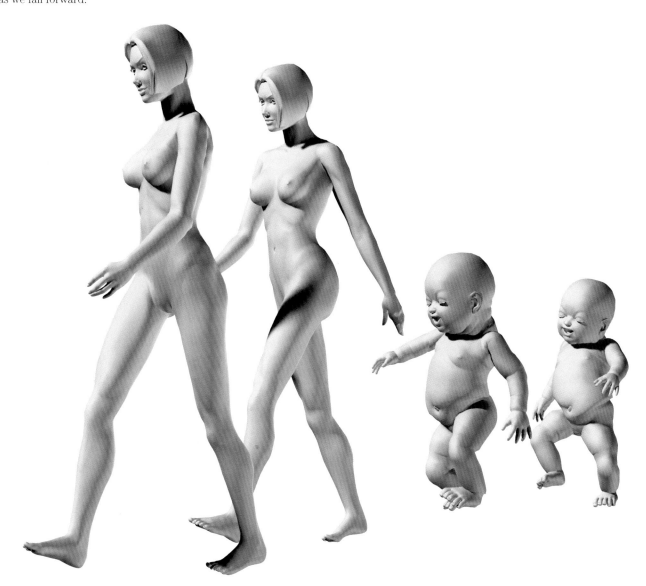

☞6. Walking and balance

Let's consider the role of the feet in walking. The foot is often thought of as a vehicle—like a tank with Caterpillar treads—that pushes the body forward. But the foot actually comes out to balance the body when the center of gravity shifts. Shift the gravity forward even more, and the

other foot comes forward. The feet carry the body naturally as the center of gravity moves.

The adult walk has been perfected and thereby lacks in variety, so consider the way a child walks. Normally, children are unbalanced and meandering. A child with a large head that is not yet stable has trouble

keeping balance in any one place. Because they wouldn't be able to sustain their weight if they put one foot ahead of the other, children tend to shuffle their feet to the left and to the right.

Keeping balance is a dynamic and important reflexive activity. The actions necessary to carry our

weight, react, and maintain our center of gravity probably have effects on many of our other movements. Understanding this goes beyond any kind of paper plans or empty theories about animation and straight to the ability to express the movement of living things in a natural way.

ranges of foot movements

In (a) we see that a foot can move in a 105-degree range, arching up 30 degrees and stretching down 75 degrees. In (b) the foot can move in a 50-degree range as viewed from the rear—25 degrees to either side. And (c) shows that the foot can move sideways in a 90-degree range, 45 degrees to both the left and the right of the center line.

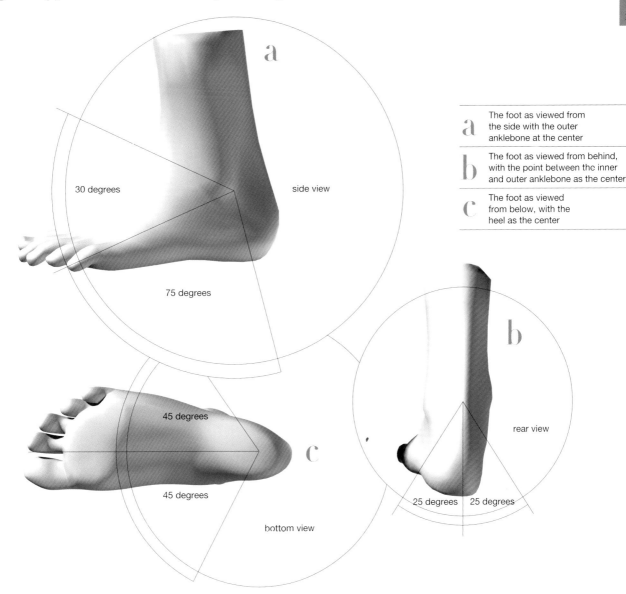

a   The foot as viewed from the side with the outer anklebone at the center

b   The foot as viewed from behind, with the point between the inner and outer anklebone as the center

c   The foot as viewed from below, with the heel as the center

30 degrees

side view

75 degrees

45 degrees

45 degrees

bottom view

rear view

25 degrees   25 degrees

Walking and the bottom of the foot

When the foot hits the ground, the toes turn upward. They touch the ground after the foot has landed. When drawing a character with very large feet, you can exaggerate the action, thereby strengthening the impact of the character's toes that are touching the ground. The foot points slightly outward when walking, with the heel on the inside. The feet are both following a center line between them. Typically, the center of gravity is on the front foot with the heel sort of floating. Shift the center of gravity to the heel, which is not made to absorb the impact, and it will hurt. Shoes absorb the impact felt by the heel, allowing us to walk for long periods. When you tire, your back slopes forward and the feet hit the ground hard. When you become even more tired, you end up just shuffling your feet along. When you get this exhausted, it's difficult to even lift your feet, and your toes begin to curl in. At that point, you are susceptible to stumbling on a bump in your path and falling.

For animators, there are two points to keep in mind. First, don't let the feet slide. To keep them from sliding, start applying the animation from the bottom. Second, the feet must depict the physical weight of the person. If you don't know where the center of gravity is, you won't be able to show this when the body is moving. And even if you spot the center, if you can't depict the feet supporting it, the picture becomes a lie.

shoulders

shoulders

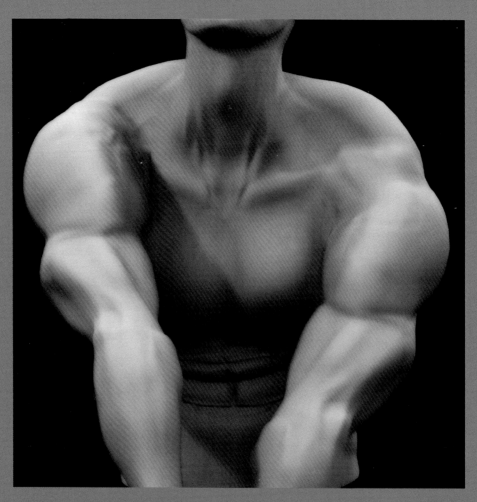

The shoulders sag, shrug, heave, express confidence, and carry burdens. They reflect a person's condition. They also can greatly change the shape of a person's body just by moving. Shoulders can endure the weight of an arm and allow it to move freely. The shoulder's ability to move in all directions helps the arm and hand move in a wider range. In fact, the shoulder, hingelike and seemingly capable of one function, has a lot of muscle, has a complex bone structure, and can exert power in all directions.

Related pages   p. 66 arms   p. 114 shapes of action
Part 2 (all) action

S
shoulders

Structure

Exploring the
Construction of
Human Body Parts

Part 1

shoulders

shoulder proportions

Example: 28-year-old male
Each collarbone is about 6 inches long. It is 7 inches from the edge of the shoulder to below the shoulder blade. And it is another 6 inches from the bottom of the shoulder blade to the top. The bone is about 4 inches across.

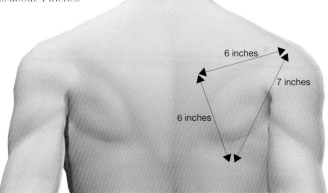

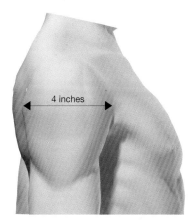

shoulder surfaces

When exposed to sunlight, the shoulder can quickly burn. A trace of hair on the surface grows toward the back. Below the armpit, you can find curly hair and a large artery. The collarbone and shoulder blade both jut out enough to be visible. Since there are no major muscles between the two, the area looks hollowed out.

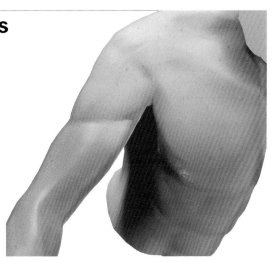

Understanding the shoulder's construction

The shoulder has the most range of any body part; it can change appearances suddenly. It plays a very prominent role in a full-length photograph, for example. Because of this, understanding the shoulder's construction is very important when creating characters for animation. Draw the shoulder with care, and your character will have more expression.

☞1. ☞2. **Most movements draw out the intimate relationship between muscle and bone**

The collarbone starts just above the first rib and extends back to the shoulder blade. The shoulder blade is triangular and connected to the humerus. The base of the humerus looks like a half ball sticking out at a slant. You can probably picture the way this joint moves, but that motion alone does not allow you to raise your arm into the air. The shoulder is flexible because of the difficult-to-grasp relations between the work of the shoulder blade and the support of the collarbone. The structure is complex—muscles also work closely with the bones.

The shoulder muscles work to allow the following movements: raising or lowering the arms to a horizontal position; lowering and raising the shoulder blade; and pulling the shoulder blades close together. Turn the page for more on the bones and muscles of the shoulder.

shoulder bones ↜1

The shoulder is made up of the humerus in the arm, the triangular shoulder blade in the back, and the collarbone, which is connected to the sternum in the center of the ribs.

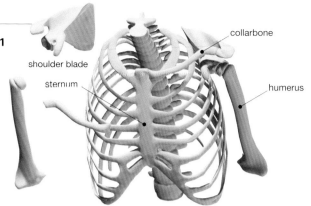

collarbone

shoulder blade

sternum

humerus

shoulder muscles ↜2

The shoulder, which can move quite freely, has all sorts of muscles that establish close connections with other body parts.

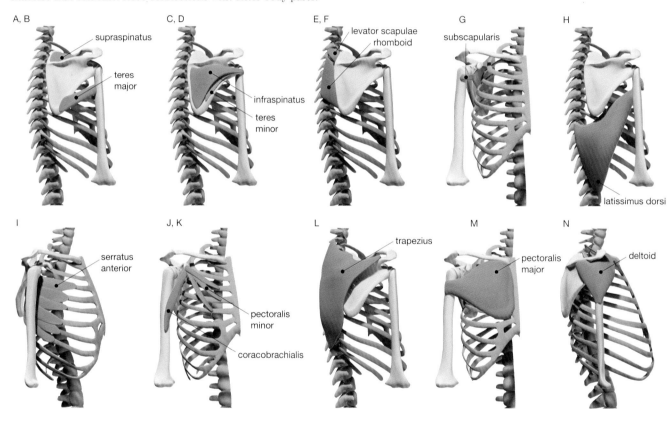

A, B — supraspinatus, teres major

C, D — infraspinatus, teres minor

E, F — levator scapulae, rhomboid

G — subscapularis

H — latissimus dorsi

I — serratus anterior

J, K — pectoralis minor, coracobrachialis

L — trapezius

M — pectoralis major

N — deltoid

↜2. Shoulder muscles and movement

Here are explanations of how the shoulder muscles displayed above help with movement.

A, B/ The supraspinatus connects along the top of the humerus and shoulder blade. It plays a role in the arm's abduction. The teres major connects the lower shoulder blade and humerus. It plays a role in the arm's adduction and involution.

C, D/ The infraspinatus and teres minor, which connect the middle of the shoulder blade and the humerus, play a role in the arm's abduction.

E, F/ The levator scapulae connects the top of the back of the shoulder blade with the neck bones. It helps the shoulder bone rise. The rhomboid connects the bottom of the back of the shoulder blade with the spine. It helps the shoulder blade pull in.

G/ The subscapularis connects the middle of the back of the shoulder blade with the humerus. It plays a role in the arm's adduction and involution.

H/ The latissimus dorsi connects the humerus and the lower spine. It plays a role in the arm's adduction and involution.

I/ The serratus anterior connects the back of the shoulder blade and the ribs. It moves the shoulder blade forward.

J, K/ The pectoralis minor connects the front of the shoulder blade and the ribs. The coracobrachialis connects the front of the shoulder blade and the humerus. They stabilize the shoulder blade.

L/ The trapezius covers a lot of the back and stretches to the collarbone. It plays a comprehensive role in the shoulder blade's movements.

M/ The pectoralis major connects the collarbone, ribs, and humerus. It plays a role in the arm's adduction and involution.

N/ The deltoid is a triangular muscle on the outside of the humerus. It plays a role in bending the shoulder joint, extending, abduction, and horizontal adduction and abduction.

S
shoulders

Structure

Exploring the
Construction of
Human Body Parts

Part 1

shoulders

pectoral arch; triangular suspension ☛3

Looking at the shoulder from above, we can see a triangular arch made by the base of the humerus, the shoulder blade supporting it from behind, and the collarbone in front. The positioning of this arch indicates what sort of movement the shoulder is making.

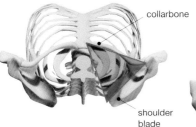

collarbone

shoulder blade

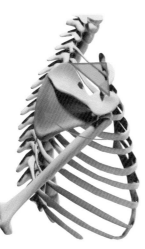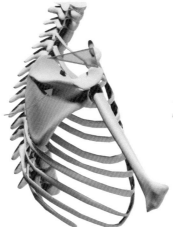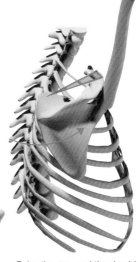

Extend the arm backward, and the shoulder blade sags a little, pointing the triangle downward.

Extend the arm forward, and the shoulder blade slides forward and up, as does the triangle.

Raise the arm, and the shoulder blade slides quite a bit forward, lifting and shrinking the triangle.

shoulder movements and armpits ☛4

Lift your arm, and the armpit tilts forward. Extend your arm straight in front of you, and the armpit points diagonally down. As the shoulder rotates, the armpit rises.

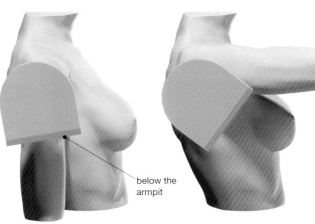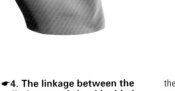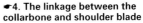

below the armpit

☛3. The triangular arch that expresses the shoulder's movement

Don't forget the triangular arch when depicting the shoulder's movement. As it alters shape, it indicates a variety of movements. By focusing solely on the shoulder joint and its mobile parts, we can't grasp the shape of the shoulder or the boundaries of its movements. By looking at the triangle, however, we can see that the shoulder is often moving forward or up. When you

consider the shape of the shoulder blade, you can see how it is hard to move the shoulder to the back. When you sleep without a pillow, your cheek and shoulder will touch. Your head slopes down, of course, but it is also clear that the shoulder moves up and in. The collarbone is easily broken. There is little muscle to protect it—it looks like a splint. The shoulder blade, with its connection to a wealth of muscles, helps create movement at the joint and supports the arm in its movements.

☛4. The linkage between the collarbone and shoulder blade

Lift your arm to a 90-degree angle, and the shoulder blade and collarbone don't have to move too dramatically. The triangle we spoke of earlier also stays horizontal for the most part. Lift your straight arm beyond 90 degrees, and the collarbone moves way back while the shoulder blade slants upward. Move the arm backward, and the collarbone pushes forward a bit while the shoulder blade inches closer to

the center of the body. Move the arm forward again, and the shoulder blade slides along in support. The connection between the collarbone and shoulder blade depicts the cyclic nature of their movements—this is the key to understanding the shoulder.

39

shoulder postures 5

The shoulder changes into various forms depending on a person's spiritual condition, lifestyle, and actions. It's also important to grasp the differences between men's and women's shoulders.

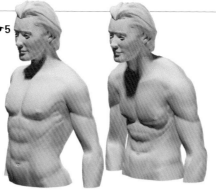

A Normal posture (left), stooping man (right)

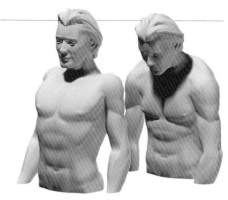

B Preparing to exhale (left), depressed man exhaling (right)

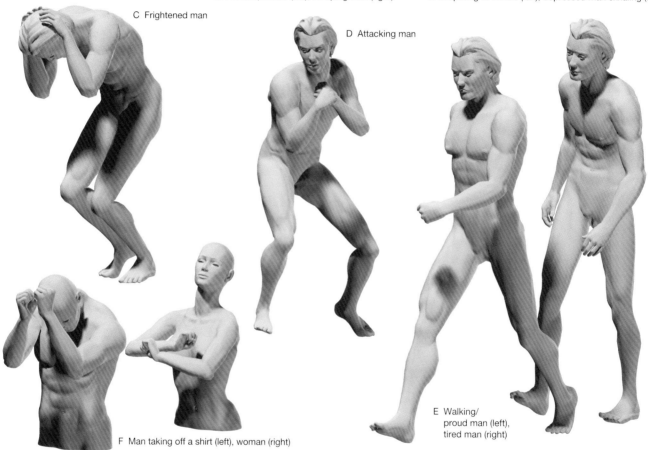

C Frightened man

D Attacking man

E Walking/
   proud man (left),
   tired man (right)

F Man taking off a shirt (left), woman (right)

5. Shoulder postures

The shoulders can unintentionally reveal something about a person's habits and spiritual condition. Let's take a look at how the shoulders do this:

A/ Relaxing

The shoulders usually sag. When you stoop, the shoulders move in and down. When your chest swells, the shoulders move back slightly as if they're leaning on the spine.

B/ Disappointment

The shoulders rise just before they sag. That makes the sagging action look bigger.

C/ Fear

When you're afraid of something, the defense instinct takes over and your shoulders reflexively cower. When you face fear, your whole body becomes rounder as you protect your head, stomach, and other vulnerable areas.

D/ Attacking

The shoulders are sturdy and easy to put one's weight behind, so they're good for charging into something, as in a rugby tackle.

E/ Walking

The shoulders can change the impression of someone walking. If the shoulders are drooping and the back is rounded, the person looks tired and weak. If the shoulders are raised and the chest is out, the person looks confident and energetic. When walking, the shoulders move in rhythm with the arms. When running, the shoulders are out first. They help the body twist. When running, you have to thrust your shoulder out as

the leg goes forward, or you'll fall. By twisting the hips and the upper body in opposite directions, you maintain balance. Swing your shoulders hard, and you can hit the ground with more force, propelling yourself forward.

F/ Putting on or taking off clothes

When putting something on, your right hand (or left, if you are left-handed) goes through the sleeve first. Then your right hand reaches back to grab the clothes item and pull it over your head, putting it on your body as

S
shoulders

Structure

Exploring the
Construction of
Human Body Parts

Part 1

shoulders

ranges of shoulder movements

The rotation shown in (a) is complicated. The arm, hanging limp, can extend horizontally to about 90 degrees. From there, with a twist of the humerus, the arm can extend another 115 degrees. It can also extend the opposite way for about 25 degrees. In (b), the arm can extend 50 degrees to the back and 150 degrees to the front, for a total of 200 degrees.

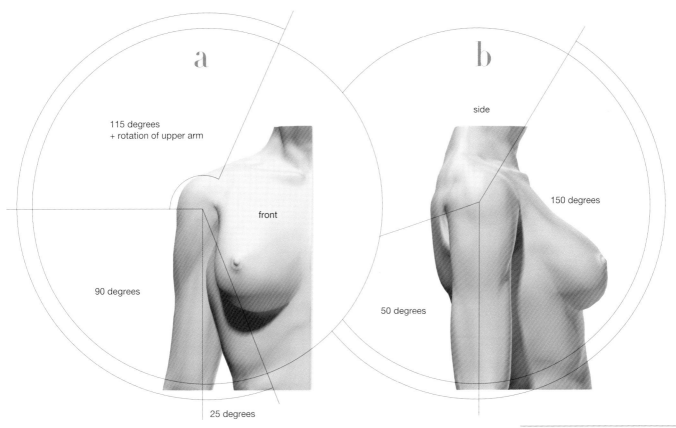

a

115 degrees
+ rotation of upper arm

front

90 degrees

25 degrees

b

side

150 degrees

50 degrees

a	The body as viewed from the front, with the base of the shoulder as the center
b	The body as viewed from the side, with the base of the shoulder as the center

your left arm goes through the other sleeve. When you've pulled your head through, your rounded shoulders stretch out and the clothes drop over your body as both hands pull the clothes. Then you may reach for the hem in the back and pull to the sides, straighten out the hem in the front, and grab the material on both shoulders to smooth out the clothing's surface. Of course, people perform this task differently, but on the whole, men and women put on clothes this way. However, the sexes

behave differently when taking something off.

Women cross their arms, grab the clothing on either side, and wedge their elbows into the material. Then they make a half turn with both arms, reach up, and take off their clothing. Men grab the clothing around the neck with both hands and pull it over their heads. Then they grab the material at the back of the neck with their thumbs, and—in one motion—extend their arms and take off their clothing. It is said that the sexes take

off their clothing in different ways because of the shape of their shoulders.

Children usually take their right hand and grab the left sleeve, pulling out the left arm. The left arm reaches from inside the clothes to the right sleeve, and the right arm is freed. Then they extend both arms down and slowly wriggle free by raising their arms, eventually taking off their tops. Children do it this way because the clothing tends to be big on them, and they don't worry about stretching

or ripping the clothes.

The question for animators isn't whether men and women really take off their clothes this way, but what approach works when depicting a feminine or masculine way of taking off one's clothes? Take care to express this difference between men and women, and you will bring individuality to the movements of your characters.

neck

neck

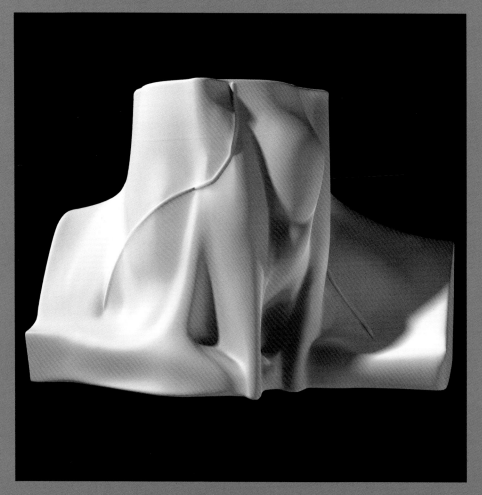

There are three important points to consider when depicting human action. The first is how to depict autonomous movement. The neck plays a special role here. People prepare to move, then they move. Imagine a person opening a door: the initiative is taken by the arm opening the door. You absolutely can't portray it in a way that looks like the door is pulling the arm. The second point is that bone and muscle structures are restricted by gravity and mechanical elements. The neck supports the heavy head and is subject to many physical restrictions. The final point is how to depict habitual actions and reflexive actions. By mastering these three points, we can create characters that feel alive. Let's investigate the role of the neck in all of this.

Related pages   p. 24 eyes   p. 72 back & waist   p. 78 head   p. 108 figure   p. 114 shapes of action

n
neck

Structure

Exploring the
Construction of
Human Body Parts

Part 1

neck

neck proportions

Example: 28-year-old male
Men's necks measure about
4.7 inches long, wide, and across.
Women's are thinner, at about 4
inches. Women's necks seem
longer because of the clothes
they wear and the way their
shoulders slope.

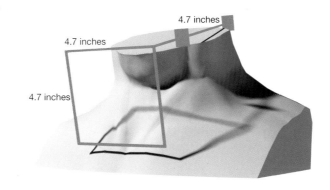

neck bones ☜1

The neck starts at the base of
the skull. Of the 24 vertebrae
(the spinal column) that extend
all the way to the sacrum, the top
seven (the cervical vertebrae) are
the neck bones.

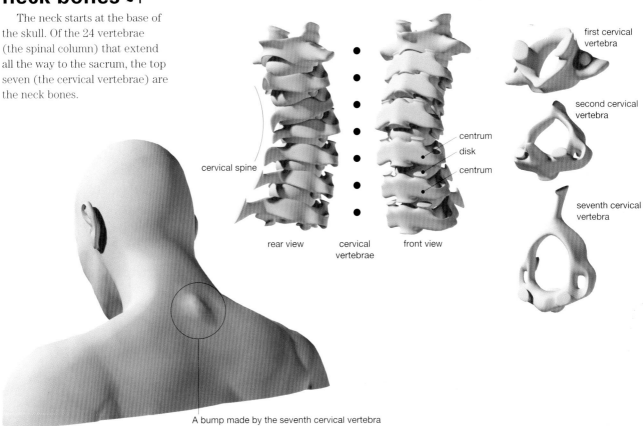

A bump made by the seventh cervical vertebra

The role of the head's pedestal—the neck

The neck is a pedestal for the head to be placed on. It plays a lot of important roles: it's the place where bundles of nerves—which transmit signals from the brain—are concentrated; it's the home of the esophagus; and it's home to the vocal cords. Also, the neck constantly supports and moves the head.

But that's not all. The neck connects the head and the torso, which means it contains many important organs. Another characteristic of the neck is that the layer of epidermis is thin, and the surface can change drastically. When drawing it, be careful not to make it look like a tube or make the head look like it is moving independently.

Also, the neck should reflect gender and age differences. If you want your animation to reflect the constant flux that is characteristic of living things, you can't overlook these points.

☜1. The bones that compose the neck and make it move

The first, second, and seventh cervical vertebrae have slightly different shapes. The other four are similar in shape. The seventh one, known as the vertebra prominens, bulges out in the back more than the others, making it visible. Because the spinal column absorbs the impact to the body, it undulates when viewed from the side. The upper part of the spine, the cervical spine, is in the neck. A newborn baby doesn't have this cervical spine yet; the head isn't fixed in position, so it totters back and forth. Once a baby turns three months old, the cervical spine is present and the head is more stable.

neck muscles

The neck contains four groupings of muscles: the subcutaneous muscles along the nape of the neck, the middle scalenes, the anterior scalenes, and the posterior scalenes. The subcutaneous muscles along the nape from the jaw to the base of the neck are protected by the platysma. The sternocleidomastoid extends from behind the ear to the top of the manubrium; it helps rotate and raise the head. One side expands when the head turns one way, and both sides expand when the head looks up.

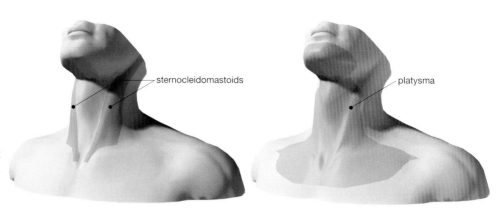

sternocleidomastoids

platysma

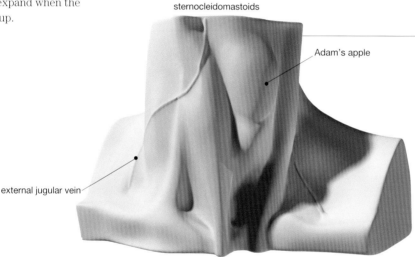

sternocleidomastoids

Adam's apple

external jugular vein

neck surfaces ☞2

The neck is an easy area to accentuate individual differences, starting with a man's Adam's apple and including the aging of the muscles and blood vessels.

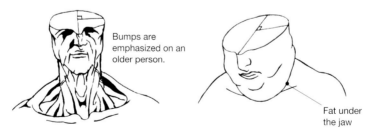

Bumps are emphasized on an older person.

Fat under the jaw

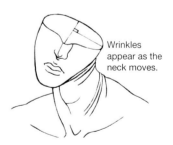

Wrinkles appear as the neck moves.

☞2. The surface of the neck contains many individual differences

A man's neck has a bumpy surface because of the large sternocleidomastoids and Adam's apple; it leaves a linear impression. A woman's neck, on the other hand, has few bumps and is thin and supple. Also, men grow hair on their neck that spreads to the back.

As for age differences, an older person's neck has less subcutaneous fat, and the muscles and blood vessels are seen clearly, giving the neck a complex contour. At the base of the neck, the collarbone and sternocleidomastoids create a cavity. The external jugular vein, which cuts diagonally across the sternocleidomastoids, can be seen on the surface when someone shouts or

exerts him- or herself. To help depict the neck well, picture the nerves, bones, muscles, and blood vessels inside the neck as being wrapped in a thin skin. In other words, it is necessary to depict the two sternocleidomastoids that move the head and the other elements in the neck like the esophagus, the nerves, and the throat.

The front of the neck has space; the

side is firm. When the neck acts, especially when it draws in the jaw or rotates, the extra skin becomes more and more wrinkled. Heavy people may have fat under their jaw, and their neck will gradually be hidden until they can barely be made out at all. Also, don't forget the shadow that a jaw casts when jutting out. The neck is a body part that clearly stands out in key light, fill light, and rim light.

n
neck

Structure

Exploring the
Construction of
Human Body Parts

Part 1

neck

neck movements ☜3

The neck's movements become the head's movements. The actual movements start with the cervical vertebrae bending smoothly and transforming the head and neck. By looking at a model of the neck, it is easier to understand the movements.

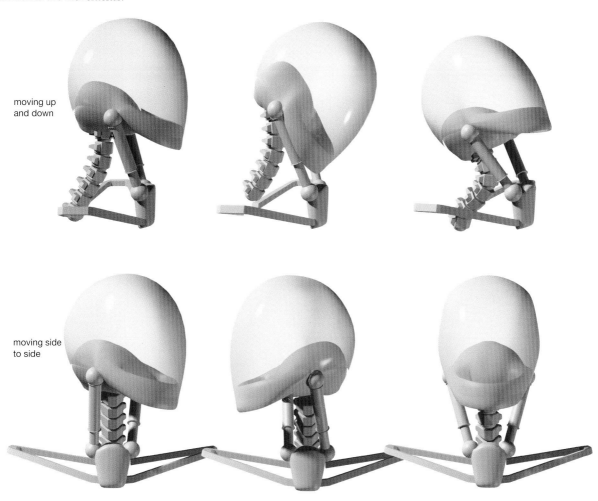

moving up
and down

moving side
to side

Throat movements

The throat moves frequently when you chew, speak, cough, sneeze, or yawn. It is said that straightening your neck and moving your jaw can help you breathe better. Another special characteristic of the throat is the way it moves slightly up and down when a person laughs.

Tension in the neck

The neck is completely tranformed when someone is enraged or excited. The changes on an elderly person's neck are especially remarkable. All of the muscles and tendons are exposed, the shoulders rise, and the hollowed-out area below the sternocleidomastoids protrudes in a complicated way. When you're

angry or you're exerting yourself, basically you draw in your jaw; when you're laughing, you tend to thrust your jaw out.

☜3 Turning one's head

The way someone turns his or her head is often misunderstood. When the head turns, the back rotates and the shoulders also move slightly in the same direction. This is not

movement supported by the neck, but by each centrum on the spine.

various ways of resting the head ☞4

When you pinch yourself between the eyes with your thumb and forefinger while leaning on your elbow, you clearly look worried. When you lean on your elbow and cover your jaw with the palm of your hand, you look troubled or as if you are reading a difficult book. When the hand is not covering the mouth, the jaw is thrust forward, and the head is still propped up by the elbow, you look like you have nothing to do. Generally, men rest their head on the back of their hand (concave), while women use the palm (convex).

A   B   C

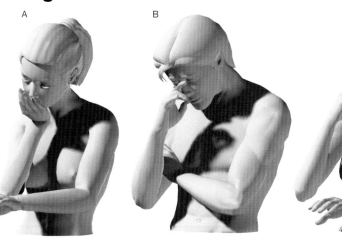

concave vs. convex postures

Men usually prop up their head with one elbow by using the back of their hand. Women use their palm or the inside of their arm. Not everyone does this, but by depicting a woman this way, she'll seem more feminine. This is a sign of gender difference that can be seen with many body parts: the man tends to use the concave and the woman the convex posture. For example, when a man looks at his watch, he'll grab his hand and look at his wrist; a woman will look at the inside of her wrist. When sitting, men prefer to sit cross-legged, while women sit with their knees together and their legs on either side. This is not primarily a factor of anatomy; rather, it comes from an aesthetic difference or one's cultural background.

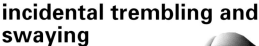

incidental trembling and swaying

Let's consider a woman drying a plate. Much of her body seems still, yet there is a continual reaction or linkage that creates movement. However, it's difficult to sense one's own trembling. Also, shaking or swaying is not completely explained by the linkage of parts. The act of breathing or the heart's beating, for example, can trigger visible reactions.

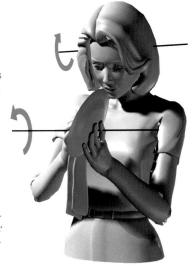

☞4. Various ways to relax on one elbow

When you're relaxed, your neck falls forward and your head is slightly raised. When sitting, the usual posture is to prop up your heavy head with one elbow. There are many poses for people leaning on their elbows, depending on what part of the head is being supported by what part of the arm or hand.

When you cover your mouth with your hand as you lean on your elbow, you look like you're thinking. Cross your arms, and the image is reinforced. This may be because you're liable to draw in your jaw. On the other hand, diagram C shows that when the hand isn't covering the mouth and the jaw is thrust out, one tends to look bored.

Gauging a person's degree of feelings or interest when he or she is scrutinizing something

If a loud noise suddenly occurs behind you, you hurriedly turn to look. If the sound is very loud, your defense instinct will reflexively make your shoulders cower and your jaw draw in as you turn. (see more on this in the "shoulders" chapter starting on p. 36.) Movements related to scrutinizing something include reflexive as well as conscious and subconscious actions.

People's eyes dart here and there

n
neck

Structure

Exploring the
Construction of
Human Body Parts

Part 1

neck

ranges of neck movements

When the head is facing forward, the neck can move 80 degrees to either side (a). This is the range of movement from the neck up; people can move more when using their body from the waist up. In (b), measuring from a line at the center of the neck and dividing the face in two, the neck can

move the head 30 degrees in either direction. In (c), the neck can move the head 65 degrees beyond the center line and 45 degrees behind it. The neck can move because of the bellowslike structure of the bones in the spine, so it really doesn't have a central spot from which rotation begins. It's

like a hose, able to bend in any direction, but if we treat the spots at the base of the neck and just below the ear and chin as the centers of rotation, we can see the different ranges. In the frontal view of (b), the point at the base of the neck can move 10 degrees in either direction. The point below

the chin has the same range. In (c), the point at the base of the neck can move 45 degrees forward and 10 degrees back. The point just below the ear has a range of 20 degrees forward and 35 degrees back.

a The head as viewed from above

b Frontal view with neck at center

c Side view with neck at center

80 degrees · 80 degrees · **a**

30 degrees · 30 degrees · 10 degrees · 10 degrees · 10 degrees · 10 degrees · **b**

45 degrees · 65 degrees · 35 degrees · 20 degrees · 10 degrees · 45 degrees · **c**

when they look at advertisements on trains. When people concentrate on something, they look at their surroundings without consciously focusing. People tend to make quick, precise movements. We spend more time waiting than acting. Because people are interested in the objects they scrutinize, we can measure their

interest by the way they look at an object. If they crane their neck toward it, they are probably very interested. But if they draw their body away from the object and pull their jaw in, they probably feel a certain abhorrence.

Scrutinizing an object is a fleeting action that adeptly expresses a

person's feelings, making it a very important part of animation production. People consider the act of seeing very important; they try to get the picture clearly reflected in their retinas. If your head is twitching, the pupils can't quite fix on a subject, so you'll steady your head. Even when your body is moving a lot, if you

are trying to see something, you'll keep your head movements to a minimum. When you shift your gaze to something else, you'll generally do so quickly. The beginning of the movement is fast; the end is slow. Unless there is a special reason, the movement starts with the eyeballs, not the neck.

chest

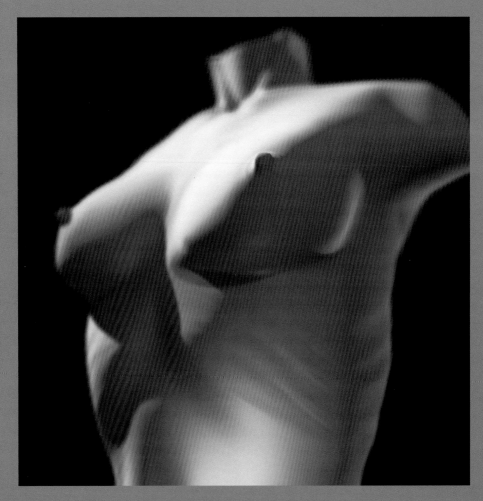

Ever since the brain was recognized as the body's central command, the image of the heart or the soul has lost some authority. But people still care deeply for the matters of the heart. The chest is made up of the heart, which pumps the blood through the body; the lungs, which control the breathing; and the ribs, which create a solid wall of protection. To put it simply, the chest is a vessel filled with some very important items. Whether or not the chest really contains the spirit or the soul, the awareness of those elements can only come through one's actions. We will explore the theme of the chest from the starting point of a young woman's average-sized breasts.

Related pages   p. 36 shoulders   p. 108 figure   p. 114 shapes of action   p. 146 running

C
chest

Structure

Exploring the
Construction of
Human Body Parts

Part 1

chest

breast proportions

Example: 25-year-old female

The chest is a body part that changes proportion, shape, and size greatly from person to person, making it hard to define a norm. We won't try to positively declare the ways to depict the breast here, but it is worth noting that if the length from the jaw to the top of the head is A, the length from the jaw to the nipple is also A.

breast geometry ☞1

The size and shape of a man's chest depends greatly on the way the muscles are formed. Women's breasts also differ greatly in size and shape. The breast itself is a lump of fat, which means it is very soft and heavy.

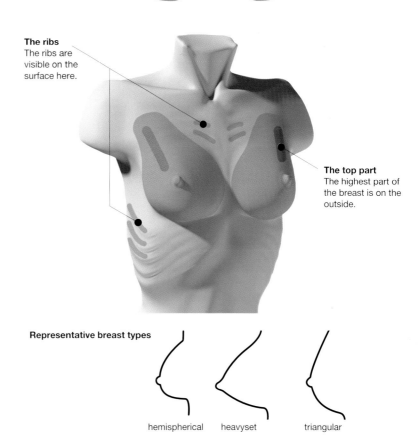

The angle of the chest
The breasts point outward.

The ribs
The ribs are visible on the surface here.

The top part
The highest part of the breast is on the outside.

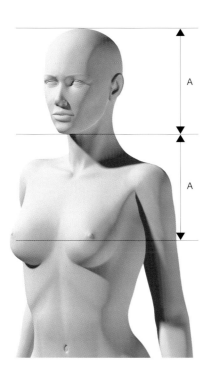

A

A

Representative breast types

hemispherical   heavyset   triangular

☞1. How gender differences affect chest sizes and shapes

There are no precise boundaries with the neck, shoulders, and stomach. Movement of the spine and shoulders can greatly alter the shape of the chest.

The male chest

A man's chest consists of muscle. Where the muscles are thick, the chest swells. Where there is little muscle, the shape of the bones can

be seen. In the center, where the muscle is not thick, ribs are visible. Along the arches on both sides, ribs can be seen angling up and back. The pectoralis major forms an L shape. The nipples are small and point out and downward. Small hairs grow in the middle and toward the pit of the stomach. The middle is also a path for sweat to travel; heat rashes and rough skin can be found here.

The female chest

Female breasts come in many shapes. We will focus on the triangular breasts common among Asians. From the side, triangular breasts consist of a half circle on the bottom and a gentle sloping line at the top that looks almost straight. Viewed from above, the breasts point outward. The nipples point in either direction when viewed from the front.

Gravity makes the breasts sag. Like a man's chest, the area between the

breasts doesn't have much flesh. The outline at the bottom of the breast is more horizontal than one would think. It starts between the nipples and extends toward the middle of the armpit. The tallest part of the breast is along the line from just below the nipple toward the collarbone. Shoulder movements play a big role in changing the breast's shape. When a woman stretches her back muscles, a side view will show the lower ribs rising.

breast bones and muscles ☞2

The chest contains the spine and the 12 curving ribs on either side. In the center is the sternum. The parts are connected by costal cartilage, except for two bones at the bottom of the ribs that are not attached by cartilage. The ribs can be seen on the surface in both men and women.

The main muscle in the chest is the pectoralis major. It can change shape greatly when flexed. The diaphragm at the bottom of the ribs and the external abdominal obliques between the ribs help in breathing. The lungs have no muscle, so they can't inhale and exhale on their own. When a person breathes, the external and internal abdominal obliques contract and relax, drawing the ribs in and out.

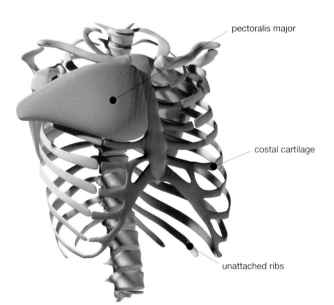

The oval-shaped ribs
Viewed from the side, the ribs look like a slightly tilted egg.

pectoralis major

costal cartilage

unattached ribs

breathing

Think of the diaphragm as a hypodermic syringe. When the rib cage expands, the lungs fill with air. When it contracts, the air is expelled.

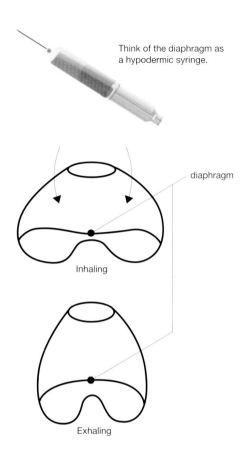

Think of the diaphragm as a hypodermic syringe.

diaphragm

Inhaling

Exhaling

☞2. The structure of breasts

The breast consists of mammary glands and fat lumps. But it is also part of a woman's sex appeal and maternal appeal because it lactates. That's why it is regularly exaggerated. Certainly, the breast is made up of beautiful curving lines. It softly bears its weight, and a woman can't control the way the lumps of fat move. The breast attracts attention.

For women, the breast is a source of pride and embarrassment, giving it a subtlety in movement. The actual movements of the breast—similar to the movements of cloth—are all reactions. The chest is wide in the base and thin at the top, forming a pyramid. When the base moves, the top of the chest shouldn't change much. It is said that the breast is not as soft as pudding and not as resilient as the devil's tongue.

C chest

Structure

Exploring the
Construction of
Human Body Parts

Part 1

chest

bras 3

The bra greatly supplements a woman's breasts. It can't be ignored when illustrating a woman's chest. The bra plays a role in the movement and outline of the breasts and the position of the nipples.

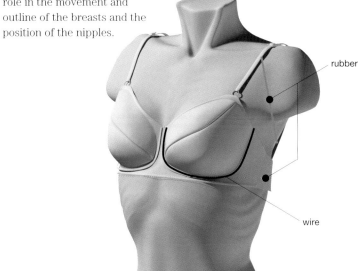

rubber

wire

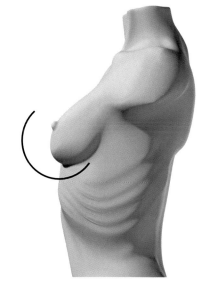

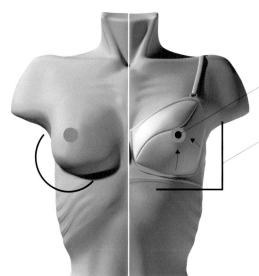

changing the position of the nipple

changing the outline

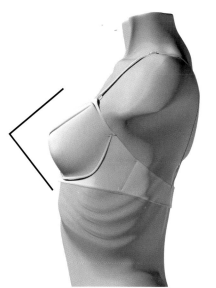

3. Bras

In most bras, a hard wire runs from the center and along the bottom of both breasts as if outlining them. Rubber is on both sides stretching toward the back to allow for elasticity. On one end, there are usually three loops that allow a woman to adjust the bra to her size;

on the other end is a little hook. The cup of the bra is shaped like the breast, and the bottom half is thick. Rubber straps extend from the top of the cup over the shoulder. The strap falls at about the point where the collarbone ends.

Bras push the breasts in and up, so the highest point of the breast

becomes higher and the tops of the breasts become closer together. The padding for both breasts is separated, and the tops of the breasts swell, emphasizing the valley between them. But unless the breasts are especially large, they are still separated.

From the side, the breasts usually

look plump and rounded, but when they are in a bra, they look straighter. Women can insert pads to make the lines more curvy.

breast movements ☞4

If you exclude the influence of the pectoralis major, the chest's movement is driven by the spine and the shoulders. Let's check out the difference between the breast's movements (its reactions to the body's movements) with and without a bra.

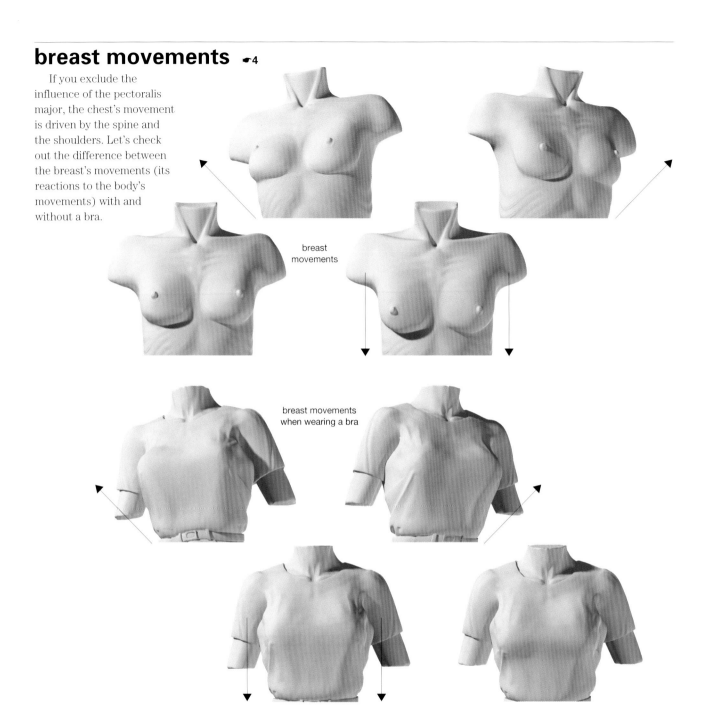

breast movements

breast movements when wearing a bra

☞4. Breast movements based on gravity and reaction

The breasts are like soft bags of material placed on top of the rib cage, following the curving lines of the ribs and hanging from both sides. If a woman bends down, her breasts hang down too; if she moves to the side, the breasts follow. They don't expand and contract. The changes we see are from chest movements and in the breast's outline.

The breasts are forced by gravity to hang down in most cases. Gravity's influence is great, and if we try to react against it, we're soon stopped. The biggest movement for the breasts is upward. They naturally hang down, so there's little room to fall farther. When breasts move side to side, they don't sway and they seem smaller. The breasts both point outward, and

lateral movement is usually caused by the body twisting. When a woman sleeps facing up, her breasts fall to the sides and her chest flattens. Reverberation after the action is slight. If she raises her arm, the breast is pushed and pulled up.

Breasts are usually encased in a bra, greatly changing their shape and movement. Most of the time, when drawing a woman, she'll be wearing

a bra, and breast movements will be restricted. The bra cups push the breasts in and up, and both breasts are more centered, creating a clear-cut image. The wire at the bottom and the adjustable strap over the shoulder turn the breast from a freely moving lump of fat into part of the chest. The breasts no longer slide to the side or hang down as much when a woman bends over.

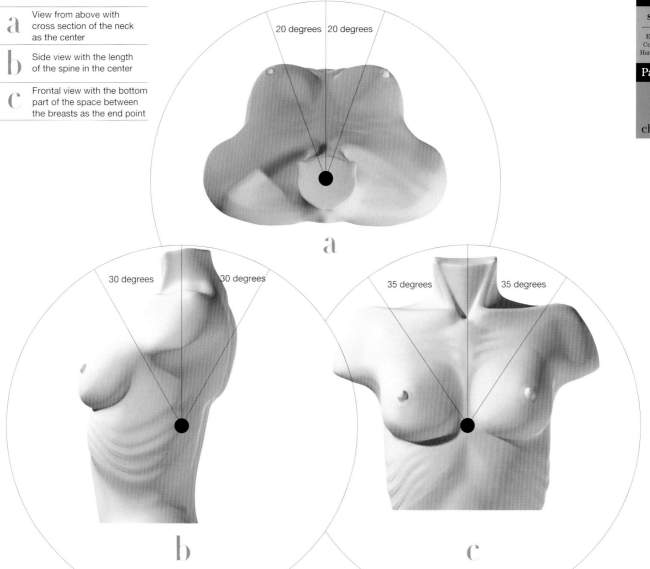

a View from above with cross section of the neck as the center

b Side view with the length of the spine in the center

c Frontal view with the bottom part of the space between the breasts as the end point

ranges of chest movements

In (a), the body faces forward. The chest has a lateral range of 20 degrees each, for a total of 40. In (b), with the body centered as much as possible on a vertical line, the range is 30 degrees either way, for a total of 60. In (c), the body is at its most vertical; the range of movement from the center is 35 degrees each way. Like the neck, the chest's range of movement is established by the spine. The use of the whole spine adds further range to the chest's rotation and movement.

Changes that occur when running

People spend very little time on the ground when running. The feet push off in a hurry, and the chest drops for a split second. From the front, the breasts make a V shape as they rise slightly after the body. If the woman is wearing a bra, lateral movements are limited, and the V shape is thinner. The breasts only change shape slightly when going down because there isn't much room to drop from their natural position. When the body springs from the ground, the breasts press against the ribs and float upward. It's almost as if there is no gravity. The breasts gently rise, fall quickly, then gently rise again. The running woman is most likely clothed, which means the tops of the breasts are aligned, depending on the clothes. A straight horizontal line is maintained at the top, and the breasts rarely move independently. When drawing breasts, remember the restrictions caused by the bra and the thickness and hardness of the clothes.

legs

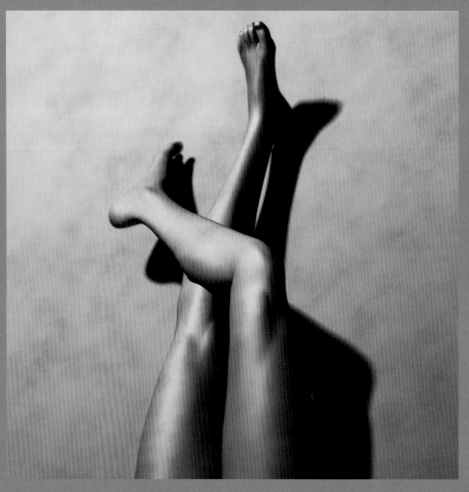

legs

Legs constantly support and transport the body. Humans, unlike many other vertebrates, walk on two legs. Because of their constant use, leg muscles are far more developed than any other in the body. Yet, they are also quick to show signs of decline. We all have some experience with aching, torn, or pulled muscles. The thigh bones, which are the largest bones in the entire body, are supported by large groups of muscles so that they can withstand great pressure and comply to the everyday demands placed on them. The Achilles' tendon, named after a mythical hero, are often used as an expression of unexpected weakness—just an example of the enormous significance legs hold for movement.

Related pages   p. 30 feet   p. 72 back & waist   p. 108 figure   p. 122 getting up

leg proportions 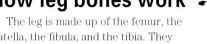 1

Example: 25-year-old female

We often see drawings of legs that are longer from the knees down than from the knees up. The fact is, the length above the knees is longer.

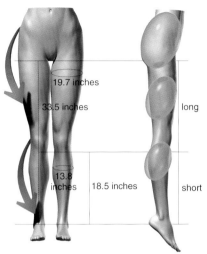

19.7 inches

33.5 inches

long

13.8 inches

18.5 inches

short

leg contours 2

Bulging muscles make the legs of men more linear, while women's legs have smoother, curving lines.

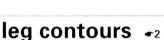

two vertical lines and horizontal lines

women

surface of knee shows complex muscle definition

men

leg structure

We use the joints at the hips and the knees to twist and turn our legs.

The hip joint allows for virtually unrestricted movement.

The knee joint has a complex structure.

The twisting of these two bones is what helps the legs rotate.

how leg bones work 3

The leg is made up of the femur, the patella, the fibula, and the tibia. They work in unison so that the legs below the knees can turn left and right.

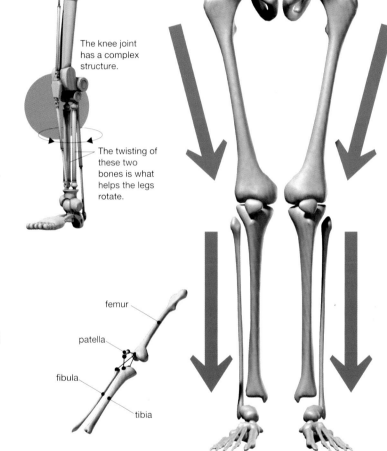

femur

patella

fibula

tibia

---

1. The three ovals—hips, thighs, and calves

Just for the sake of analyzing leg movements, we will define legs here as starting from the buttocks down to the ankle. We will also focus on the contours of the legs. The side view of the legs reveals an oval-shaped outline in the hips, thighs, and calves. Contrary to those smooth lines are the rather rough lines created by the shin and knees. The frontal view of the legs shows a gentle curve of the thighs, continuing through the calves.

As for the overall outline of the legs, the lines curve in from the top of the thighs to the knees, then slightly flare out to form the calves.

2 The surface and outline of the legs

The knee joints are visibly complicated. The part known as the kneecap consists of a small bone called the patella. When you stretch out your legs, the skin around the knees may form wrinkles. The back of the knees has two vertical ligaments

and a lot of wrinkles. The overall outline of a man's knees is muscular and tends to appear linear. The woman's legs have flowing curves. For some women who work out, the outline is muscular, especially around the calves.

3 The leg bones

The leg consists of three long bones and one small one. The femur is in the thighs. The patella, which slides when it moves, is in the knees, and the two bones below the knees

are the thick tibia and the thinner fibula. The femur is a large and strong bone. The top end, shaped like a hemisphere, is attached to the hip bone. The lower end is flat and connects with the tibia. The fibula and the tibia help create the rotating movement below the knees from left to right. The patella moves up and down with the movement of the knees, thanks to groups of muscles that tug from above and below the knees.

leg muscles ☞4

The thighs rely on their front, side, and back muscles to bend and straighten out the legs. The two muscles that form the calves are connected to the Achilles tendons.

leg surfaces ☞5

Drawing the fine and complex lines and wrinkles on the surface of the knees gives the picture more definition. Differences in skin conditions can make the area below the knee a little shiny or darkly discolored. For gender differences in skin surface, see notes.

The look of a woman's legs can change due to tanning or shaving.

rectus femoris muscle

patellar ligament

hamstrings (bulges created by the biceps femoris muscle, semitendinosus muscle, and semimembranosus muscle)

bulges created by gastrocnemius muscles

Achilles tendon

kneecap movements

The knee has a wide radius of movement. The lower end of the femur sticks out at the back, and the tibia basically rotates along the protuberance of the femure.

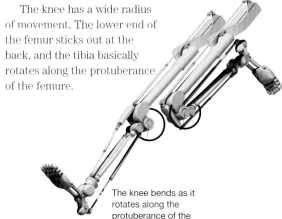

The knee bends as it rotates along the protuberance of the femur.

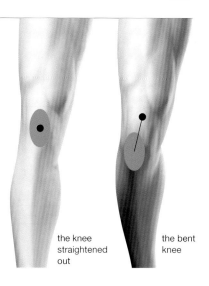

the knee straightened out

the bent knee

☞4. Leg muscles you can observe through the skin

The rectus femoris muscle, which runs diagonally across the thigh, connects the kneecap and the outer hip bone. It will flex when you lift up your leg. In the back of the thigh, the hamstrings—consisting of the biceps femoris, semitendinosus, and semimembranosus muscles—will flex. The patellar ligament connects the patella and the top of the tibia. The main thigh muscles are located

in the inner thigh. They all connect to the pelvis, femur, and tibia. The two main muscles used to bend the knee are located behind the thigh and connect the pelvis to the bones below the knees. These form the two lines visible behind the knees. The muscles needed to straighten out the legs are found in the back of the thighs and stretch from the pelvis to the patella. The main muscle groups below the knees are the ones that form the calves—the gastrocnemius

and flat soleus muscles. They connect to the Achilles tendons.

☞5. Gender differences in skin texture

There are lots of hairs on the surface of men's legs, but not on the inner thighs or ankles. The knees have lots of fine lines, and the skin is thicker and rougher compared to the rest of the legs. The main point in tightening the look of the legs is in depicting the complex ridges that are

present on the knee. Women wear skirts, so the appearance of their legs can change quickly below the knees. Women also have hair on their legs, but because they shave, the texture of their shins can appear shiny.

*l*egs

Structure

Exploring the
Construction of
Human Body Parts

Part 1

legs

supporting the body

In a standing position, the legs constantly support the body from the waist up. Naturally, muscles are more developed in the legs than elsewhere in the body. Standing upright with the heels together is tiring. Open your legs wider and the muscles in your outer thighs will tense up. The most comfortable position for your legs is opening them slightly, at shoulder's width. This brings the femur and the tibia in a straight line, strengthening vertical support. When required to stand in one position for a long period, we take the most comfortable position—standing straight with the legs slightly apart. In most cases, we also shift our upper body weight to either of the legs. We rest our upper body on one leg and when that leg gets tired, we shift our body weight onto the other leg. This is only one of many examples that illustrate this search for comfort, which is also seen in the way we toss and turn in bed.

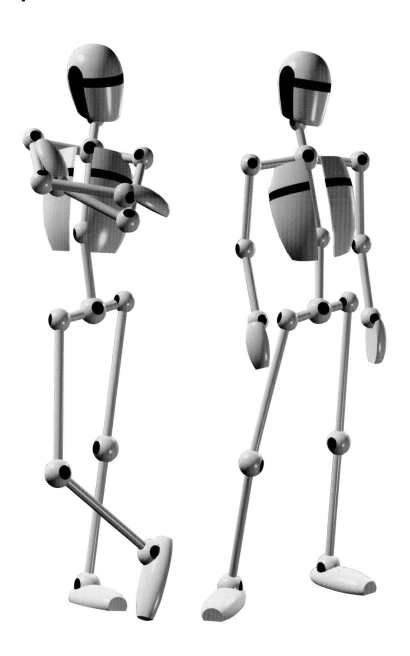

The waist as part of our legs

The area starting from the abdomen and extending below moves largely in accordance with the legs. You could say the buttocks and the outer hips are considered part of the legs. This is apparent, given the way most of the leg muscles cover a wide area of the pelvis, from which they extend. The hips swing to stay in sync with movements of the legs. Even in the simple act of standing, the hips do not remain straight.

sitting and resting ☞6

There are many ways to sit: sitting cross-legged on the floor; sitting on a chair; and sitting with your legs folded on the floor. When taking a seat on a chair, unless you are aware of it, your legs will part slightly. We cross our legs and switch them over to tense up and relax the leg muscles.

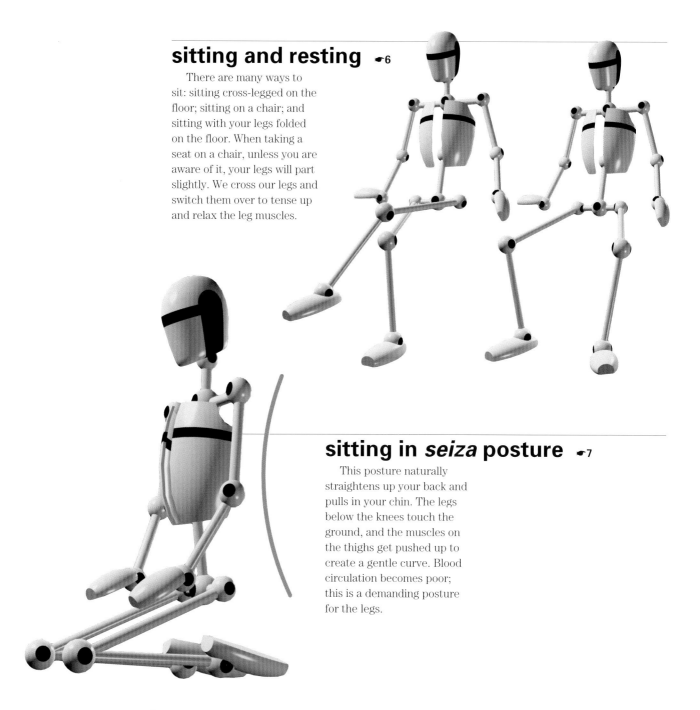

sitting in *seiza* posture ☞7

This posture naturally straightens up your back and pulls in your chin. The legs below the knees touch the ground, and the muscles on the thighs get pushed up to create a gentle curve. Blood circulation becomes poor; this is a demanding posture for the legs.

☞6 Sitting on a chair, crossing legs

When we sit on a chair, the relaxed muscles and fat expand the thighs a little on the seat. When you put on weight, the thighs become larger, thereby separating the knees more. Women sit with their legs closed because they wear skirts or want to look more elegant. Most men sit with their legs open. When they sit—just as when they stand—they put their weight on one of their legs by crossing their legs. Lifting even one

leg off of the ground means one less load to carry. It is a comfortable position that creates just the right amount of muscle tension. Fat thighs create a space right below the upper knee when the legs are crossed. Men tend to stretch out their feet as long as there are no restrictions caused by the shoes they wear. Of course, keeping the same leg crossed for a long time causes poor blood circulation and becomes uncomfortable, so they are switched regularly.

☞7 *Seiza*—a Japanese idiosyncrasy

We will look at a unique way of Japanese sitting– *seiza*. It used to be said that *seiza* led to bowleggedness. It turns out that early walking (i.e., in late infancy) is to blame. Bowleggedness is a condition in which the legs curve out at the knees when standing straight with both heels touching. The opposite condition—crossleggedness—is also well-known. *Seiza* sitting naturally straightens out the spine with

beautiful posture. It is said to be a touch of Japanese ingenuity to make the most of small living spaces. In the *seiza* position, your legs below the knees straighten out and touch the ground, while the thighs get pushed up from below to create a gentle curve. Your back straightens out and your body weight ends up resting on the tips of your feet. This causes poor blood circulation and numbness. There are some veterans who can sit in this position for hours, but for most, two hours is the limit.

ranges of leg movements

$\mathcal{l}$ legs

Structure

Exploring the
Construction of
Human Body Parts

Part 1

legs

In (a), the leg, with the foot facing straight ahead, rotates 40 degrees to the left and 40 degrees to the right—a total of 80 degrees. In (b), the foot's rotation of 25 degrees each to the left and right, totals 50 degrees. In (c), the entire length of the straightened leg can swing back 45 degrees and swing up 135 degrees, for a total sweep of 180 degrees. In (d), the axis is at the knee. With the leg resting vertically, the shin part can move only 5 degrees forward, and bend 145 degrees to the back, swinging a total of 150 degrees. In (e), shows that the legs can swing left to right like a pendulum to a total of 97 degrees—27 degrees to the inside and 70 degrees to the outside.

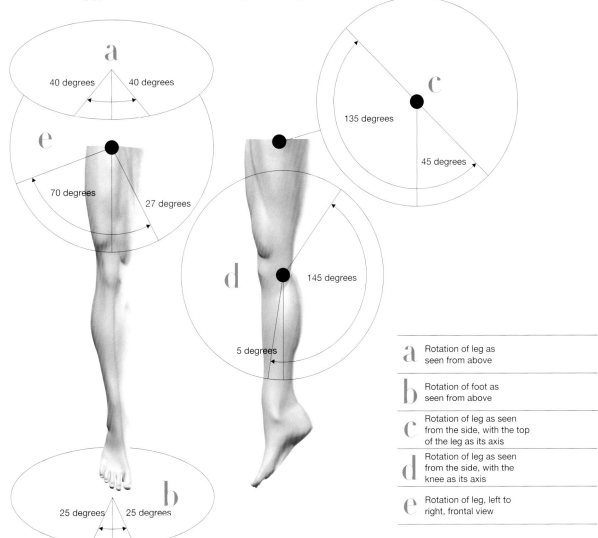

a — 40 degrees — 40 degrees

e — 70 degrees — 27 degrees

c — 135 degrees — 45 degrees

d — 145 degrees — 5 degrees

b — 25 degrees — 25 degrees

a	Rotation of leg as seen from above
b	Rotation of foot as seen from above
c	Rotation of leg as seen from the side, with the top of the leg as its axis
d	Rotation of leg as seen from the side, with the knee as its axis
e	Rotation of leg, left to right, frontal view

Habits of the legs—"the shakes"

In standing, walking, and running, the legs take on a significant burden. And yet, they get in the way of comfortable posture. As a result, they get treated pretty roughly. This is a good opportunity to bring up one such example in relation to habits of the legs. The shakes—characterized by the foot, slightly raised at the heel, making small and rapid movements up and down—is one of those habits. Some say it is a sign of minor agitation, but it also seems to happen at idle times or for other inexplicable reasons. Either way, it seems to happen when the legs have nothing to do or when they are in the way. For example, when you are sitting on the floor with your legs straight out, you may shake the tips of your foot left to right. This could be considered a shake. Other habits include standing on one knee or elevating the leg to a desk or chair. In other expressive movements of the legs, stomping is often seen at times of anger or frustration. Jumping for extreme joy, or making small and quick steps when it's cold is another example.

hands

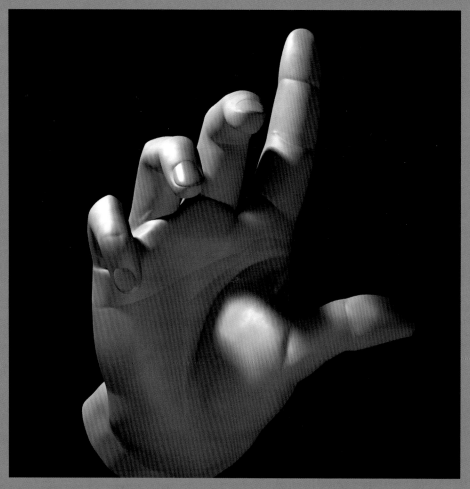

The hands are called the "second face" because they can express rich, complex emotions. They touch, grab, carry, turn, pinch, point, push, hang up, stroke—there's no end to the functions of the hand. People use their hands in all sorts of operations. They help people communicate; in fact, sign language is communication itself. The hands are all-purpose machines when it comes to making things; once they make an instrument or appliance, they can proceed to make an even larger one. The hands also help the body by protecting and attacking. In this chapter we'll investigate the hands, focusing on how they are put together and how they move so precisely and elegantly.

Related pages   p. 66 arms   p. 108 figure   p. 114 shapes of action   p. 122 getting up

h
hands

Structure

Exploring the
Construction of
Human Body Parts

Part 1

hands

hand proportions

Example: 25-year-old man
The longest point on the hand is from the wrist to the tip of the middle finger; it's about 7.9 inches. From the wrist to the tip of the thumb is just 4.7 inches. The width of a hand is rarely more than 3.1 inches.

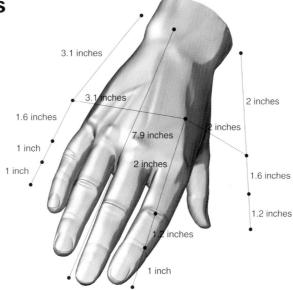

3.1 inches

3.1 inches

1.6 inches

7.9 inches

2 inches

2 inches

1 inch

2 inches

1 inch

1.6 inches

1.2 inches

1.2 inches

1 inch

hand bones 🖙1

The palm has five metacarpal bones. The thumb has two bones, and the other fingers have three each. At the base of the hand, there are eight small carpal bones.

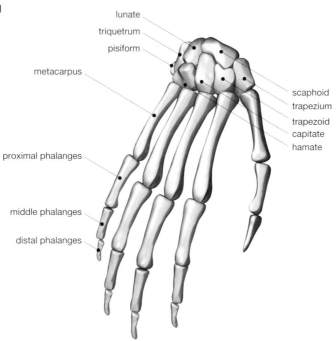

lunate

triquetrum

pisiform

metacarpus

scaphoid
trapezium
trapezoid
capitate
hamate

proximal phalanges

middle phalanges

distal phalanges

🖙1. Is the thumb really the longest?

The hand has five fingers. When thinking of movement and construction, it may be more accurate to consider the palm a finger too. The palm is often seen as a flat board; the fact that it can transform flexibly is forgotten. The palm contains the bones of the fingers as well as muscles, and though it can't move on its own, it can transform along with the fingers. The

thumb has one less joint than the other fingers, but if you consider the metacarpus at the base of the hand as the third joint, it becomes the longest finger. Next are the middle finger and the ring finger. This may seem a little odd, but because the placing of the base of the fingers is a little off, the middle finger is the only one that looks long. The length of the middle finger and the width of the palm are about equal.

🖙1. Nails are not bones; they're hardened skin

Nails protect the fingertips, but their special quality is that they reinforce the fingers' functions. The nails are there because the fingertips would be soft otherwise and unable to perform all sorts of small tasks. Thanks to the nails, we can pick up small items. Functionally, nails seem like bones, but they are actually hardened skin.

hand muscles 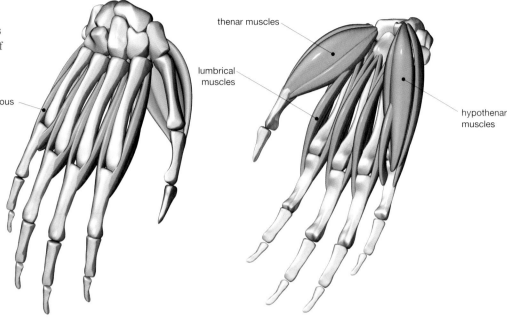2

The muscles are found mostly in the palm of the hand. They help the fingers open and close. But most of the muscles that move the fingers are in the arm.

dorsal interosseous muscles

thenar muscles

lumbrical muscles

hypothenar muscles

hand surfaces 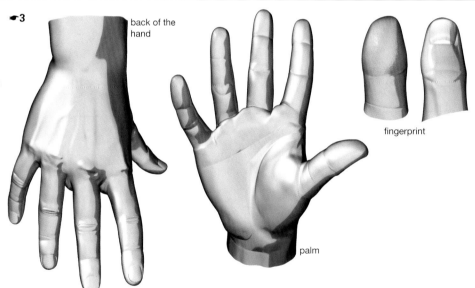3

The skin on the back of the hand is thin. Bones, tendons, blood vessels, wrinkles, and hair are all visible. The palm is very difficult to depict because of the many large lines running across it. Fat creases run across the finger joints, and the fingers have prints at the top. Prints are also on the swelling around the base of the fingers. Fingerprints change from person to person.

back of the hand

fingerprint

palm

☞2. The muscles in the palm move the fingers

Muscles are connected to movements in the metacarpus. The lumbrical, palmar interosseous, and dorsal interosseous muscles are in between the metacarpal bones; they help the fingers open and close. The opponens digiti minimi, flexor digiti minimi brevis, and abductor digiti minimi muscles are along the base of the pinky finger. These muscles are called the hypothenars. The big

muscles along the thumb, known as the thenar muscles, include the abductor pollicis brevis, flexor pollicis brevis, opponens pollicis, and adductor pollicis.

The bumps on the back of the hand change with movement

The many protuberances on the back of the hand can be difficult to depict when the hand is moving. Due to gravity, blood flows downward, swelling the blood vessels along the

surface. Typically, the veins get fatter when you let your hands drop and thinner when you raise them.

☞3. Drawing the wrinkles on the hand

The wrinkles on the back of the hand come from little laterals in the wrist and the many little muscles in the middle. Around the knuckles, the grain gets bigger and rougher; it stretches out and the skin whitens when the fingers are bent. The area

around the next joint (proximal interphalangeal joint) has lots of wrinkles, like an oval whirlpool. Like the knuckle, the wrinkles stretch, and the larger muscles and tendons become more apparent when the finger bends. The wrinkles around the final joint (the distal interphalangeal joint) are weaker. The nail, made of hardened skin, will grow steadily over the course of one's life. Tiny grooves run across the nail; at its base is a small, white arch.

h hands

Structure

Exploring the
Construction of
Human Body Parts

Part 1

hands

open hand and closed fist ☞4

The fingers move after the tendons deliver the message that the muscles in the forearm have contracted.

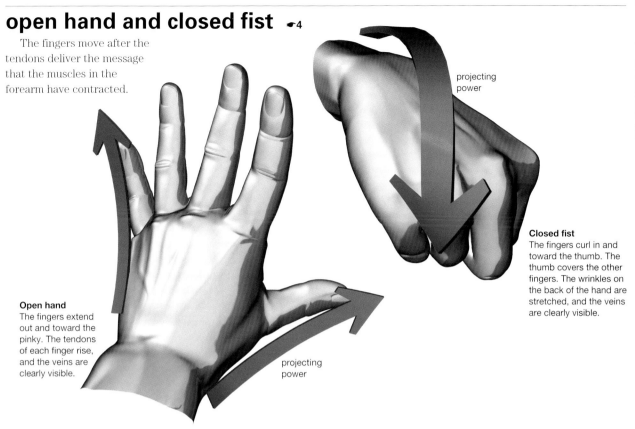

projecting
power

Open hand
The fingers extend out and toward the pinky. The tendons of each finger rise, and the veins are clearly visible.

projecting
power

Closed fist
The fingers curl in and toward the thumb. The thumb covers the other fingers. The wrinkles on the back of the hand are stretched, and the veins are clearly visible.

how hands work ☞5

The main muscles for moving the fingers are not in the hand. The muscles of the cervical spine contract when fingers move—very much like the magic hand novelty gifts. Clench your fist and the tendons in that arm will naturally tighten in succession. The pinky and the thumb are exceptions in that, unlike the other fingers, they have extra muscles in the palm of the hand to help them move. Otherwise, all five fingers move in the same way.

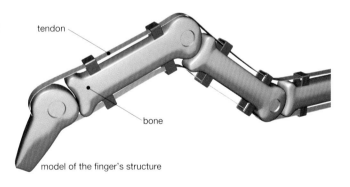

tendon

bone

model of the finger's structure

☞4. Fingers aren't straight

When you open your hand, your fingers tend to gravitate toward the pinky finger's side without your intending them to. The space between the metacarpals and the angle between each finger opens up. The first joints above the knuckles also slightly bend backward. There are wrinkles above the second joint. The skin on the back of the hand slackens, making the veins more pronounced. Two tendons in the thumb are stretched, and a triangular

hollow is made between the thumb and the index finger. Also, the slack skin between these two fingers is emphasized as it thins out. Open your hand as far as it will go, and the thumb and pinky reach farther back than the other three fingers. Power is projected along these two fingers.

☞4. Creases formed by a clenched fist

When you make a fist, your fingers tend to move in and toward the thumb. The metacarpus creates an

arch. The space between the fingers disappears and the creases on the joints stretch. The thumb covers the other fingers, and the skin between the thumb and index finger is twisted into an arc of wrinkles. Along the pinky, there is excess skin that creates wrinkles heading toward the inside. Clench the fist, and the wrist turns toward the inside, delineating the bundle of tendons in the middle. Because the pinky slides toward the middle, distorted wrinkles appear between each finger. The power is

projected from the back of the hand to the end of the pinky finger.

☞5. The joints above the knuckle move in sync

Most of a finger's movements are linked to the wrist. Moreover, the first and second joints on the fingers move in sync. They can't move independently.

63

hand postures ☞6

The most basic action of the hand is grasping something. Even newborn babies will try to grab objects near them. It's practically a reflex.

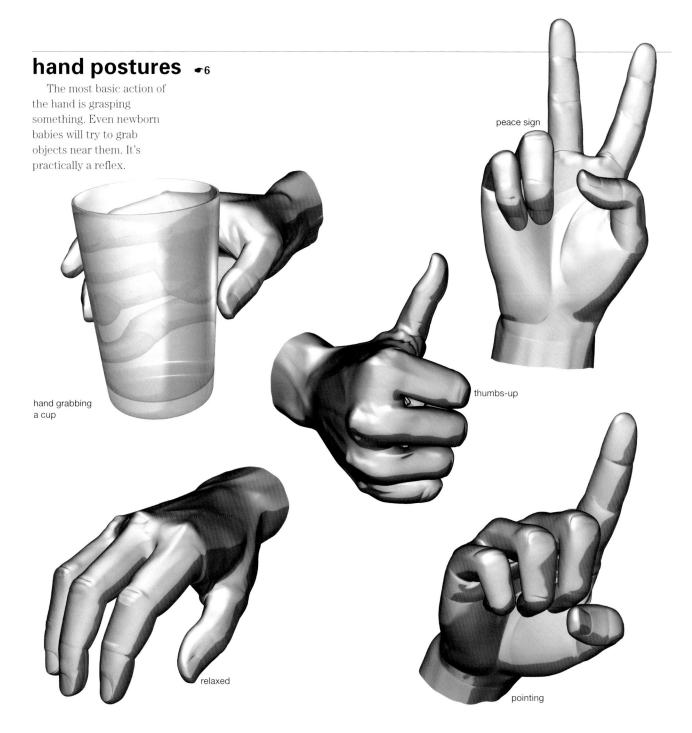

hand grabbing a cup

peace sign

thumbs-up

relaxed

pointing

☞6. Grasping, throwing, catching

Babies become aware of objects and instinctively try to grab them when they are just two or three months old. Next, the act of throwing is quickly learned. But the skills needed to catch something take much longer to acquire. The difficulty in playing catch is that one has to gauge the placement, speed, and mass of the object to prepare for the impact. The hand absorbing the impact also can't just drop the ball; it has to grab it and throw it back. This action also shows how much and how cleverly the hand and brain are linked.

☞6. Grabbing a cup and lifting it to one's mouth

There are all sorts of ways to grab things. The fingertips are loaded with nerves, and they can tell with a touch the nature of something, including how thick it is. For example, when grabbing a cup on a table and bringing it to your mouth, you'll take the following steps. First, your hand will grab the cup. The fingertips immediately report to the brain that the cup is thick and can be grabbed firmly without breaking, or that the surface is smooth and cold. If you know beforehand that the cup is filled with water and heavy, the fingertips will exert strength when touching it. After grabbing it firmly, the fingertips will estimate the strength needed to lift the cup. Once the lifting begins, the fingerprints will act like the grooves in a tire, keeping the cup from slipping. The hand lifts the cup to the place it estimates the mouth to be while also trying not to spill. The important thing is not to estimate the weight of the object while lifting; the fingertips will do this before the lifting begins. The hands can handle these difficult actions with no waste and with elegance.

h
hands

Structure

Exploring the
Construction of
Human Body Parts

Part 1

hands

ranges of hand movements

From the index finger to the pinky, the joints at the bases of the fingers can rotate outwards about 10 degrees each. The thumb and the skin connecting it to the palm work in a way that allows the thumb to make big movements. One point to remember: the wrist can't move along the axis formed by a vertical line extending through the middle finger; that movement is controlled by the upper arm.

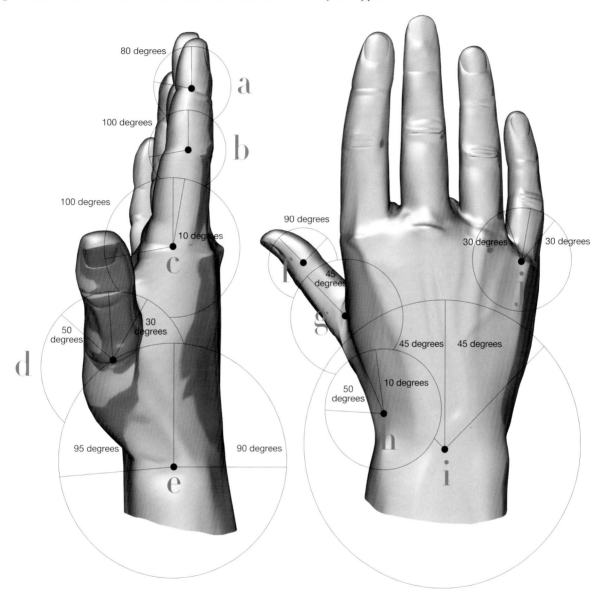

☛6. The role of the hand, master of the elaborate move

The hands express emotion. We can applaud the many gestures of the hand—they are almost like speech. Hand gestures are heavily influenced by culture, so they aren't quite a universal language. But the hands play an important role in communication and expressing emotions and desires. If you want something, you extend your arm to take it. If you're bored, you may tap your fingers on your desk. When you want to put your thoughts down on paper, your hand grabs a pen and writes. The hand is capable of elaborate moves: just think of a musician's fingers flitting over an instrument. The hand is second only to the eyes in expressing our desires. When people shake hands, the meaning is clear because it's the hand that is expressing the spirit's intent.

 Side view of the hand, with the axis at the center

 Bird's-eye view of the hand, with many axes

✳ The wrist is not capable of rotating by itself.

arms

arms

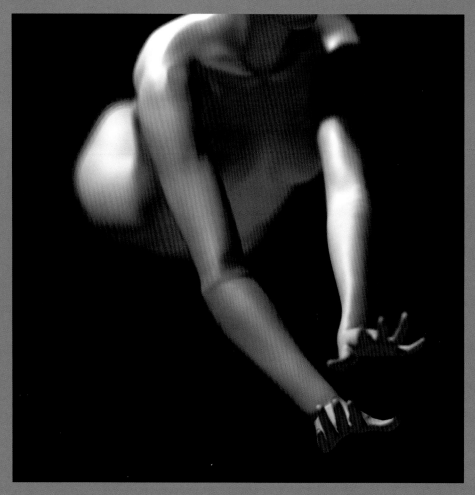

We deal with the shoulders, arms, and hands in separate chapters, but they are intimately related. In this chapter, we will look at all of these parts in a comprehensive explanation of the arms. In the "hands" chapter, we only touch on the fact that the forearm contains the muscles for moving the fingers. The arms, though not as strong as the legs, are capable of supporting the body. Compared to the legs, with their abundance of continuous power, the arms excel by having sudden bursts of power. At first glance, the arms seem simply constructed, but just like the legs, they have a network of muscles that let them make complex movements. The arms can do everything from lifting heavy barbells to pantomime or sleight of hand.

Related pages   p. 36 shoulders   p. 60 hands   p. 72 back & waist   p. 108 figure   p. 114 shapes of action
Part 2 (all) action

a

arms

Structure

Exploring the
Construction of
Human Body Parts

Part 1

arms

arm proportions

Example: 28-year-old male

From the top of the shoulder to the elbow is about 12 inches. From the elbow to the wrist is about 10 inches. The base of the arm is about 4 inches wide; the elbow is 3.5 inches; and the wrist is 2.8 inches.

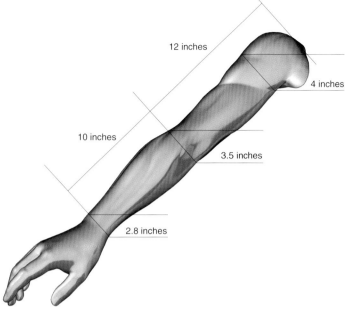

12 inches

4 inches

10 inches

3.5 inches

2.8 inches

arm bones ☞1

The arm has three bones: The humerus, which connects to the shoulder; and the ulna and radius, which connect to the wrist.

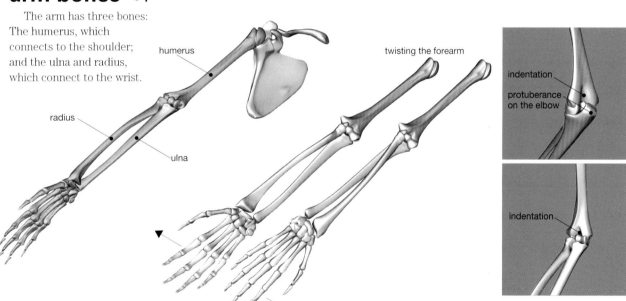

humerus

twisting the forearm

radius

ulna

indentation

protuberance
on the elbow

indentation

☞1. The special shape of the three arm joints

The humerus, in the upper arm, connects to the shoulder. It has a half-circle-shaped projection at the place where it connects with the shoulder blade. This big half circle helps the shoulder make big moves. The humerus is the biggest bone in the arm; the part that has flattened near the elbow is the epicondyle. It

has three important protuberances on it that connect with the cavities on the ulna and radius. Also, the protuberances on these forearm bones settle in the cavities on both sides of the middle of the elbow joint.

The bones in the forearm help the arm bend and extend. Extend your palm out in front of you, and the ulna will be on the inside, the radius on the outside. The ulna starts out thick and

bends slightly at the end. When you bend your elbow, the shape of the ulna can be made out as it pushes out. The radius is thick at the bottom and bends out slightly. When these bones twist, the forearm turns and the wrist rotates with it. The wrist joint can't turn by itself. The three bones and joints of the arm all have special characteristics.

arm muscles ☞2

The arm shows the muscles more than any other body part. The muscles that bulge the most are the deltoids, which cover the shoulders; the biceps brachii; and the triceps brachii. The other muscles are listed in the diagram.

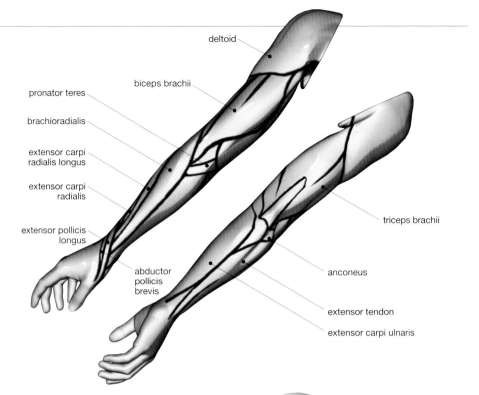

deltoid
biceps brachii
pronator teres
brachioradialis
extensor carpi radialis longus
extensor carpi radialis
extensor pollicis longus
abductor pollicis brevis
triceps brachii
anconeus
extensor tendon
extensor carpi ulnaris

arm surfaces

Pale hair grows here, and sometimes veins are visible. Extend your arm and the extra skin around the elbow turns into a mound of fine transversal lines. The skin on the tip of the elbow and around the wrist is thin and hard. The skin on the elbow is a little thick and rough. Also, depending on the clothes one wears, the skin may become tanned in places.

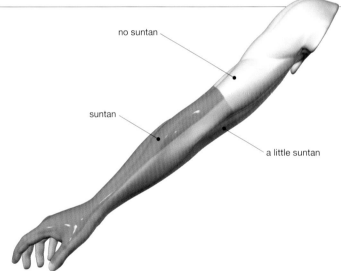

no suntan
suntan
a little suntan

☞2. The muscles responsible for the prominent bulges

The deltoid covers the shoulder. The pectoralis major in the chest is also a shoulder muscle; the arm can make it transform. The slightly depressed area where the deltoid and pectoralis major come together creates a groove. In the back, the infraspinatus, teres minor, and teres major all stretch from the shoulder blade to the humerus. The upper arm has five muscles; three bulge visibly. Wrapped around the back of the humerus is the triceps brachii; the biceps brachii is in front. The triceps brachii is so named because it splits into three. It starts above the elbow at the ulna and extends to the armpit and the shoulder blade. The biceps brachii, which splits into two, starts at the radius and extends to the shoulder blade. It can grow quite a bit; when the arm bends at the elbow, it distinctly bulges. The muscle starts on the inside and heads slightly to the outside. In the inside of the upper arm, the coracobrachialis extends from the middle of the humerus to just before the bulging part of the shoulder blade, but it doesn't show on the surface.

The brachialis is diagonally below the biceps brachii on the other side of the arm. It connects the middle of the humerus to the top of the ulna. A complex array of muscles that control finger movements run from the elbow to the forearm. The pronator teres goes from the bottom of the humerus to the top of the radius. The brachioradialis and the extensor carpi radialis longus and brevis are found in the bundle of muscles that run along the outside of the arm from the elbow to the wrist. These muscles create the biggest bulges in the forearm. Along the thumb are the abductor pollicis longus, extensor pollicis brevis, and flexor pollicis longus. The other fingers are controlled by the extensor tendon, which splits into four on the back of the hand; the flexor digitorum superficialis in the palm; the flexor carpi ulnaris; and the palmaris longus, which connects with the membranous tendons fanned out across the palm. Other muscles in the forearm include the supinator and the pronator quadratus.

68

a
arms

Structure

Exploring the
Construction of
Human Body Parts

Part 1

arms

arm movements ☞3

The arm moves in the shape of a fan, with the shoulder as the center. The shoulder may be considered ball-shaped in its movements, but when you consider gravity and the average range of movement, it seems more fanlike. The starting point of the fan becomes blurred because of the range of movement of the collarbone and shoulder blade. The range of the elbow is a fan or a half circle. Also, remember the arm's ability to rotate. A simple move by the arm quickly becomes complex when the shoulder and forearm rotate.

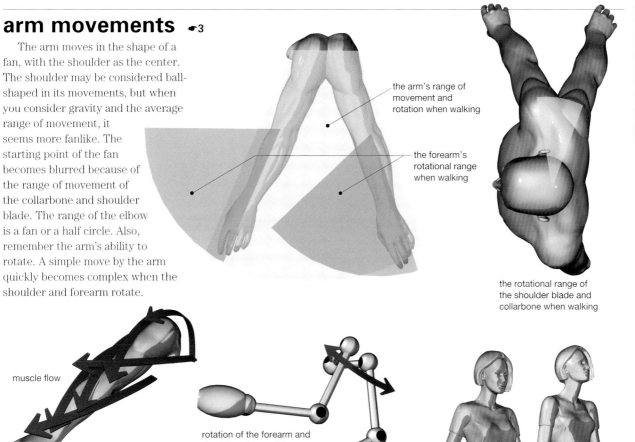

the arm's range of movement and rotation when walking

the forearm's rotational range when walking

the rotational range of the shoulder blade and collarbone when walking

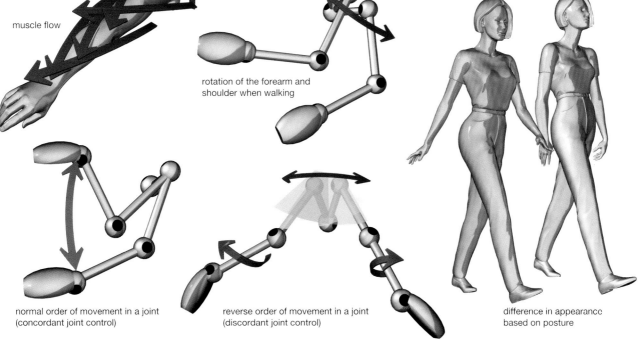

muscle flow

rotation of the forearm and shoulder when walking

normal order of movement in a joint (concordant joint control)

reverse order of movement in a joint (discordant joint control)

difference in appearance based on posture

☞3. Blood vessels change shape as the arm is raised and lowered

Some people have arms with very apparent veins around the inside of the joints and in the forearm. The arm hangs down usually, and the blood flows that way, thickening the veins at the bottom. Raise your hand, and the blood thins out, leaving the veins difficult to see. The elbow changes the most with movement, but the forearm can also change when the bones inside are twisting.

☞3. Concordant and discordant joint control

When the arm is conscious of something in the hand, it moves differently. When the hand is trying to get something, the starting point for its movement is known as discordant joint control. When you walk or do something else where the arms move naturally, the movement starting in the shoulder is known as concordant joint control. All the body parts follow the movement of the adjacent part closest to the waist. For example, the hand follows the movement of the forearm; the forearm follows the upper arm; the upper arm follows the torso. Discordant joint control is when a body part moves against this flow.

When walking, the way the arms swing and chest moves hints at a person's character. These movements reveal individual differences and special characteristics. A person who walks briskly with rapid arm swings, the chest out, and the fingers extended gives off a very different impression from a person who walks with the back slouched, the arms dangling, and the fingers closed in a fist.

arms as balancers 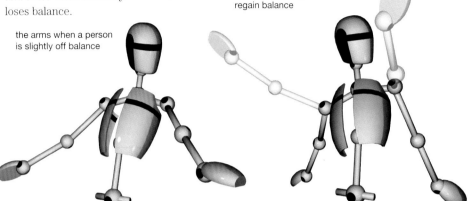 ◄4

Arms can help balance by adjusting their own weight and shifting the center of gravity. The arms widen out and twist when the body loses balance.

the arms when a person is slightly off balance

the arms rotating to regain balance

the arms going into a protective posture as a person loses balance

arms as shields ◄5

The arms play all sorts of roles: they transport, search, attack, communicate—but most of all, they protect the face.

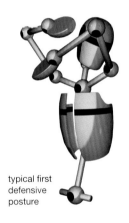

typical first defensive posture

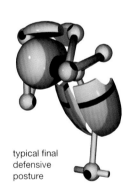

typical final defensive posture

age/sex differences and left/right dominance

When a woman wants to look thinner, she tries to improve her posture and appearance by making her arms and legs look thinner and more beautiful. A "third leg" to old people is an arm and a cane. They put the other arm behind their waist. For many people, the size and development of the left and right arms differ. Because we favor one arm over the other—such as when we play sports like tennis, where one arm does most of the work—the difference in the arms is plain to see.

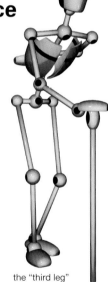

the "third leg"

◄4. Two attempts at maintaining balance

To maintain balance, you can shift the weight of your arms in order to shift your center of gravity, or you can try to regain your balance by rotating your arms. The latter attempt is more out of desperation. Normally, when you cannot maintain balance by moving your arm, you try to grab onto something nearby. When that doesn't work, you rotate your arm as you try to recover your balance.

When even that doesn't work, you quickly extend your arm in the direction of your fall to create a cushioning effect. When you fall, your arm or leg may end up getting injured. In video games where the characters fight, this aspect is often eliminated; the characters fall in an unnatural way.

Imagine standing in a bumpy train without holding onto a strap. The arms become useful—not in protection, but in building momentum.

The arms are also helpful in the broad jump when trying to elevate oneself, or in ice skating when trying to quickly spin. When landing in the broad jump, people send their arms back in a big motion to keep from falling backward.

◄5. The most important role of the arms: protection

People instinctively put a priority on protection over aggression. We react to even trivial difficulties with quick

precision. The arm instantaneously goes right in front of the head to protect it. Even if it's just a loud noise that surprises us, we hunch our shoulders. Also, when we step out into bright sunlight or water is spraying at us or a strong wind is blowing our way, we put both our hands up with the palms out and cross them to protect the face. When an even bigger danger is near, we crouch and clasp our hands behind our necks.

a

arms

Structure

Exploring the
Construction of
Human Body Parts

Part 1

arms

ranges of arm movements

Looking from above in (a), the arm can move 100 degrees to the inside and 50 degrees to the outside for a total of 150 degrees. In (b), with the thumb extended, the arm can rotate 45 degrees in either direction. In (c), with the upper arm directly down, the arm can move 170 degrees to the inside.

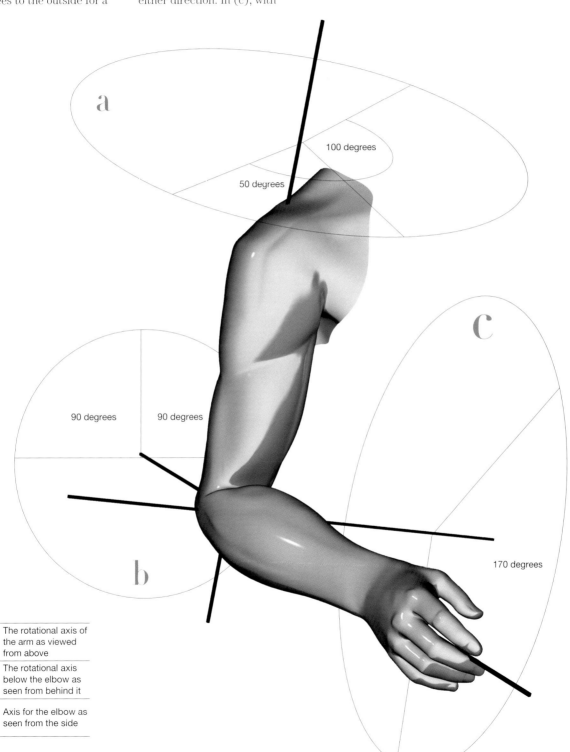

a   100 degrees

  50 degrees

b   90 degrees   90 degrees

c   170 degrees

a	The rotational axis of the arm as viewed from above
b	The rotational axis below the elbow as seen from behind it
c	Axis for the elbow as seen from the side

back & waist

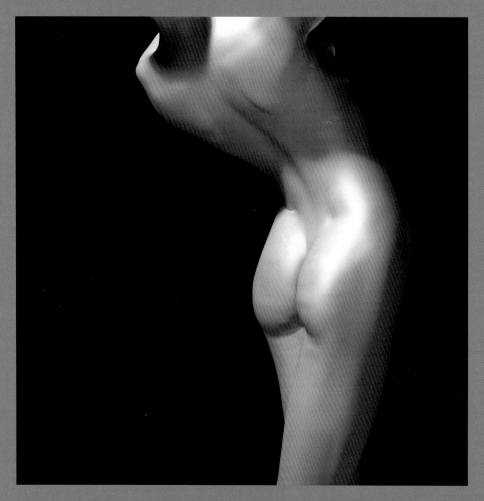

The waist is essential for bodies that stand erect. It's where the body's actions start; the rest of the body parts follow its lead. The waist is a hard worker—it helps the legs rotate and the body bend. In contrast, the navel at the top part of the waist moves very little. The waist is the body's center, and if the center of gravity should slip far from here, a person will fall. The back is also the body's pillar of support. It has all of the elements necessary for people to stand. The back connects the waist to the head; it also supports the shoulders and arms, while the ribs protect the organs.

Related pages   p. 30 feet   p. 36 shoulders   p. 42 neck   p. 48 chest   p. 54 legs   p. 60 hands
p. 66 arms   p. 78 head   p. 108 figure   p. 114 shapes of action   Part 2 (all) action

b & w
back & waist

Structure

Exploring the
Construction of
Human Body Parts

Part 1

**back &
waist**

hip proportions

Example: 25-year-old female

The length from the bottom of the buttocks to the base of the neck is about 21 inches, and to the beginning of the rib cage is 7 inches. The hip bone is about 11 inches across at the top.

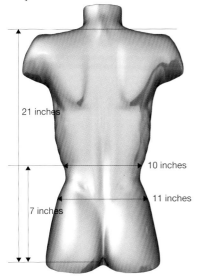

21 inches

10 inches

11 inches

7 inches

back bones ☜1

The spine is made up of 24 vertebrae. They are divided into the cervical, thoracic, and lumbar regions, moving from top to bottom. The spine connects with the pelvis, an area where the difference between men and women is great.

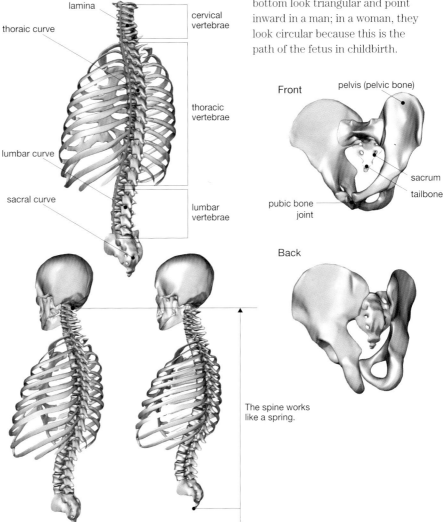

lamina

thoraic curve

cervical vertebrae

thoracic vertebrae

lumbar curve

sacral curve

lumbar vertebrae

The spine works like a spring.

pelvis shapes ☜2

The pelvis is divided into the pelvic bone, which spreads to either side, the sacrum in the middle, and the tailbone below that. The pelvic bone, shaped like an open cylinder, is different in men and women. The holes at the bottom look triangular and point inward in a man; in a woman, they look circular because this is the path of the fetus in childbirth.

Front

pelvis (pelvic bone)

sacrum

tailbone

pubic bone joint

Back

The waist is the key to depicting the human body

If you can vividly capture the movement of the waist, you will succeed in depicting the human body. If you can depict the waist, you'll be able to transmit to your audience the softness and suppleness of the body.

☜1. Construction of the back

The spine consists of 24 vertebrae. The top seven are called cervical

vertebrae. They bend forward. The next 12 are called thoracic vertebrae, and they bend back. The ribs start here. The remaining five are called lumbar vertebrae, which curve forward. The vertebrae have a complex shape with many bumps. They get thicker going from the cervical to the thoracic regions; they thin out going toward the lumbar area. Each vertebra has two lateral tubercles, a thorny tubercle on the bottom, a nodular upper joint above,

and, directly opposite, a nodular lower joint. Some of these protuberances can be spotted when a person bends his or her back; they are also bunched along the back of the neck. Each vertebra has little range of movement on its own, but the whole spine can greatly transform the body. The curve in the spine supports the weight of the head and protects the body from the impact of a blow.

☜2. Constructing the pelvis: the pelvic bone, sacrum, and tailbone

The pelvic bone comes together at the bottom, buffered by cartilage. The sacrum in the middle has many dents on the surface. Beyond that is the tailbone. Thanks to the process of evolution, we no longer need this bone; it's a remnant of the days when we had tails. To support the torso and keep us standing on two legs, the bones in our waist are bigger than those of four-legged animals.

back/hip muscles 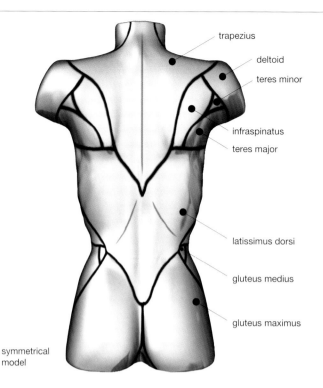3

A lot of muscles have to function together so that the spine can move in a flexible way. There are muscles that connect the waist and back; muscles that connect the waist to the femur in the leg; and muscles in the buttocks.

trapezius

deltoid

teres minor

infraspinatus

teres major

latissimus dorsi

gluteus medius

gluteus maximus

symmetrical model

abdominal muscles 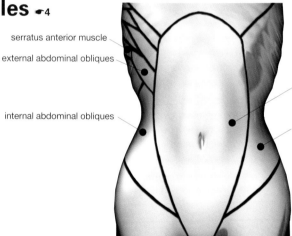4

The rectus abdominis muscles cover the abdomen. The muscles begin at the center of the lower ribs and extend all the way to the pubic bone in the pelvis. The external abdominal obliques are on both sides; the internal abdominal obliques, which connect the rib cage with the pelvis, are farther in. The transversus abdominis muscles are even farther in.

serratus anterior muscle

external abdominal obliques

internal abdominal obliques

rectus abdominis muscles

transversus abdominis (bottom layer)

☛3. Some muscles show their bulges; some don't

The iliocostalis lumborum, longissimus thoracis, and spinalis thoracis are all large muscles found deep in the body starting from the sacrum and around each vertebra. The muscles that connect the head and the spine include the splenius capitis, longissimus capitis, longus colli, longissimus cervicis, and splenius cervicis. In the middle are the longissimus thoracis and

iliocostalis thoracis. These muscles are in the middle of the back, but none of them can be seen on the surface. Short muscles in the neck include the semispinalis and multifidi. Some very short muscles that connect with nearby vertebrae include the transverse process muscle, interspinales, and rotators. This is all covered by the latissimus dorsi and the trapezius, which spread from both sides of the spine. These muscles show on the surface; it's

important to firmly draw their points of contact.

☛4. Muscles in the waist

First, the quadratus lumborum connects the pelvis and spine. Its chief role is to connect the waist with the femur in the leg. The big muscle on the buttocks is the gluteus maximus. The gluteus medius and the gluteus minimus are on both sides of the front of the abdomen. They chiefly deal with moving the legs to the

outside and fixing their position. The muscles extending from the tailbone to the femur include the pubic muscle, adductor longus, adductor magnus, adductor brevis, and gracilis. They chiefly move the leg to the inside. The iliacus muscle connects the femur, spine, and pelvis, and extends and contracts the waist. A lot of other muscles extend to the knees. Almost none of these can be seen on the surface, with the exception of the gluteus maximus and medius.

b & w
back & waist

Structure

Exploring the
Construction of
Human Body Parts

Part 1

back &
waist

back/hip surfaces ☞5

The indentations in the back of the waist and the ridges formed by the muscles on the spine should not be overlooked. Through training, the muscles on the abdomen can stand out, but usually there is a lot of fat there, and the muscles can't be seen.

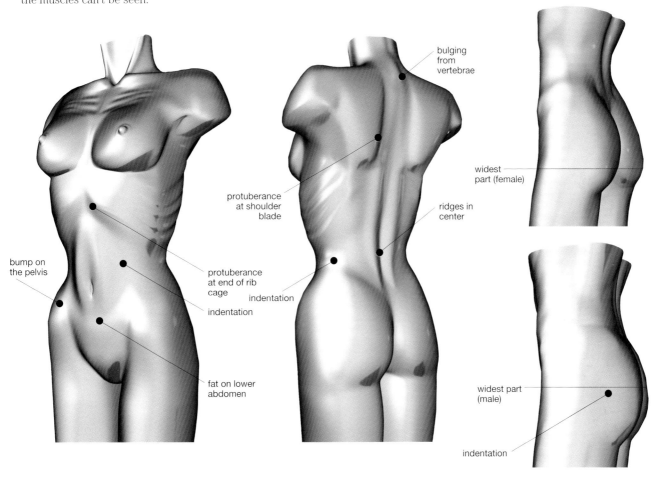

bump on
the pelvis

protuberance
at shoulder
blade

protuberance
at end of rib
cage

indentation

indentation

fat on lower
abdomen

bulging
from
vertebrae

ridges in
center

widest
part (female)

widest part
(male)

indentation

☞5. Bulges and indentations on the waist, back, and buttocks

The back has all sorts of bulges and indentations. Check the diagrams above for details on the indentations at the back of the waist or the bulges of the muscles along the spine. Also, pay attention to the smooth ridge at the end of the rib cage, the indentations along the navel, and the half-circle knoll on the lower abdomen. Bumps from the pelvic bone can be seen on both sides of the waist, with small indentations running along the back. The shape of the buttocks is very different for men and women.

A woman's waist is vertically longer and the widest part is at the bottom. A man has big indentations on either side of his waist, and the widest area is higher. The abdomen also varies greatly, depending on body type. The end of the rib cage is more pronounced on thin people; the lower abdomen stands out on fat people. The constricted part of the waist is on about the same level as the navel, in between the ribs and the abdominal muscles. A sharp indentation runs exactly down the middle from the chest to the navel. This area differs greatly from person to person. Remember that there is no bone in the abdomen—it's important to bring out its softness.

back/hip movements ←6

The back is the central pillar of the body, and the hips are the foundation. Once people began standing on two legs, the waist took on more importance.

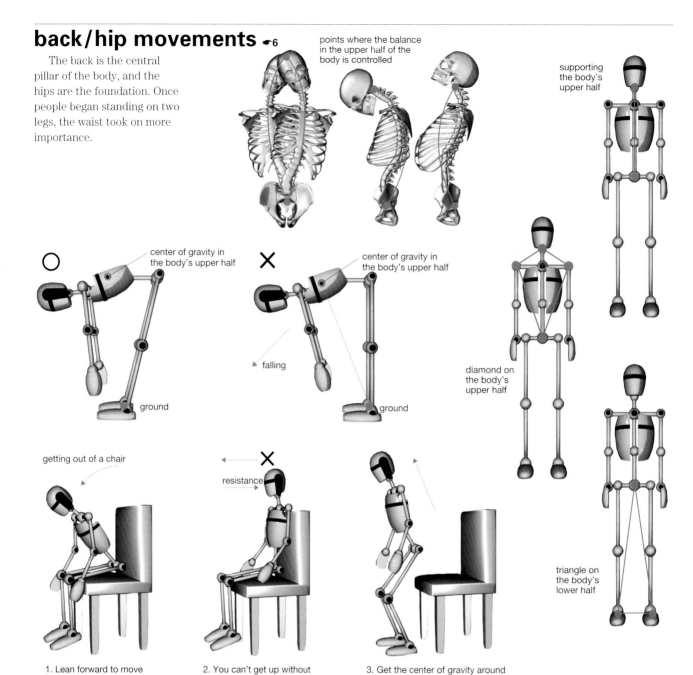

points where the balance in the upper half of the body is controlled

supporting the body's upper half

○ center of gravity in the body's upper half

× center of gravity in the body's upper half

falling

ground

ground

diamond on the body's upper half

getting out of a chair

× resistance

triangle on the body's lower half

1. Lean forward to move the center of gravity.

2. You can't get up without leaning forward.

3. Get the center of gravity around your waist and stand.

←6. Balancing the upper and lower halves of the body

When both feet are on the ground, you form a triangle with your lower half, which is indispensable to balancing. The line between the waist and the base of the neck is important for balancing the upper half of the body. Add the shoulders, and the diamond shape between the waist, shoulders, and neck is formed. The base of the neck and the waist at the legs move in nearly opposite directions. If they moved in unison, you'd fall over. Keeping balance is probably the waist's most important task. If you ignore this subconscious ability to adjust the body's balance, you won't be able to create lifelike animation. Be careful to depict the flexing of the back and the tilt of the neck. With even a simple turn of the head, the angle of the back can change, moving the diamond.

Think of the back as a cylinder, and you'll see that it doesn't clearly tilt or rotate in different directions. The back is very detailed and can bend quickly in any direction. But if the waist is set and there is no other support for the upper half of the body, the diamond's movement is restricted. Bend your back, and your waist will naturally adjust. To prove this, consider how you bend when standing. If you think that you just bend the upper half of the body and keep the lower half steady, you're wrong. Actually, as you bend, your hips push backward. If they didn't, you'd fall. Watch from the side as a person bends over, and you'll see that the shoulders—not the waist—are at a right angle.

Protecting the center of gravity is an assignment levied to the waist and back; moving that center is their mission. When the back and waist can't keep balance, the arms help; they can control the center of gravity by moving in a variety of ways.

ranges of back/hip movements ☞7

b & w
back & waist

Structure

Exploring the Construction of Human Body Parts

Part 1

back & waist

In (a), the body faces forward; the back and hips can move 70 degrees to either side, for a total of 140 degrees. In (b), the upper half of the body moves 50 degrees to either side of the center line, for a total of 100 degrees. In (c), the upper half bends backward 60 degrees and forward 150 degrees. The spine has a bellowslike structure, and with the exception of (a), there are no special central points where rotation occurs.

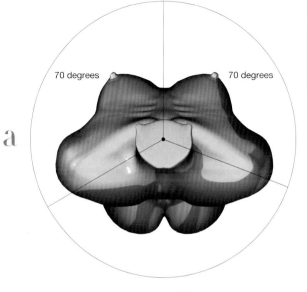

70 degrees · 70 degrees

a

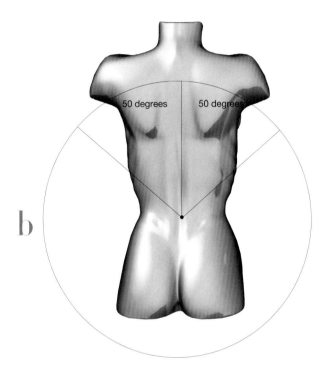

50 degrees · 50 degrees

b

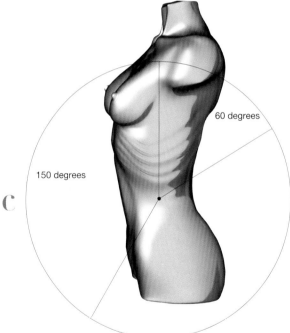

60 degrees

150 degrees

c

a — The rotational axis of the body as viewed from above

b — The navel as the central point, viewed from behind

c — A horizontal line through the navel as the central point, viewed from the side

☞6. Getting up from a chair

Probably everyone has tried this once: you use one finger to keep someone from getting out of a chair. Press one finger against the head of a person sitting in a chair, and that person can't get up. The theory is simple, but it clearly expresses the relationship between the waist and the back. The fact that you can't stand up unless you lean forward is something that needs to be experienced; until then, the one-finger trick feels like magic. The waist, which protects the center of gravity, can release a lot of power if the spine can flexibly and efficiently move that center.

☞7. The pliant back and its limited range of movement

The back has lots of small mobile parts and muscles, allowing it to bend flexibly and freely. But it carries the weight of the upper half of the body, and doing that while keeping the body centered restricts its movements. The back rarely moves independently. The spine viewed from the front tapers off at the top like a diamond. The center of gravity around the abdomen is the thickest part of the body, and from there, it basically becomes more flexible. The ribs begin at the vertebrae in the back, so when the vertebrae transform, the ribs also bend. Suddenly, the whole rib cage will change.

The waist is influenced by the legs, changing its position or rotation every time they move. Take a look at the diagram showing the changes that take place in the spine's thoracic curve through to the line in the lower abdomen. The soft and hard parts in the waist are easily identified. Make sure to be very clear on where the pelvis ends.

head

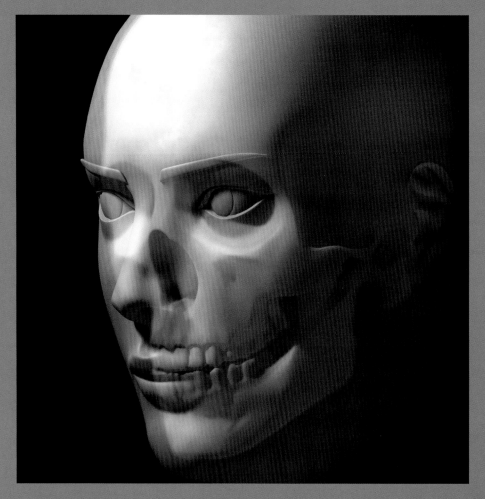

The head is the body's command central. The skull is the armor that protects the head. The skull also determines the shape of the face. It may seem like one big bone, but it is actually 22 bones that—with the exception of the jawbone—are fused together. Gender and age differences are clearly reflected in the skull. Over the years, age differences become particularly clear, especially when you consider that the size of our brain and eyeballs stays the same throughout our lives. But the skull isn't mobile; the head is moved by the neck. The only mobile part of the head is the jawbone. Let's look closer at the shape and structure of the head, especially the skull.

Related pages   p. 24 eyes   p. 42 neck   p. 72 back & waist   p. 90 nose & ears   p. 102 face
p. 108 figure   p. 114 shapes of action   p. 170 looking back

h
head

Structure

Exploring the
Construction of
Human Body Parts

Part 1

head

skull proportions

Example: 28-year-old male

From a side view, the skull fits into a square. The top of the skull is about 6 inches across. The length and width are both about 10 inches.

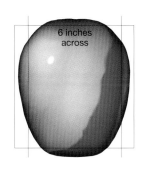

6 inches across

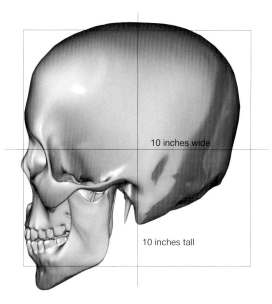

10 inches wide

10 inches tall

skulls ☜1

The skull has been used for a long time in paintings and images as a symbol of death. It is the most structurally expressive set of bones in the body.

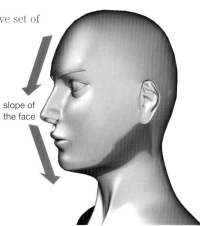

slope of the face

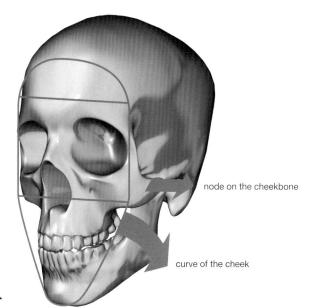

node on the cheekbone

curve of the cheek

☜1. Investigating the contours of the face

Take time to notice the hard and soft places of the face, as well as its bumps and indentations. It's difficult to properly capture the contours of the face because we tend to concentrate on the most prominent characteristics—the eyes, nose, and mouth. Picture the skull with a layer of skin covering it. First, draw a square. Divide it with a line down the center, fitting the face, including the jaw, in one half. Next, a horizontal line down the middle should pass through the eye sockets. Most of the head will fit into the upper half, while the teeth and jawbone go in the lower half. Our jaws are said to have stopped developing because we eat soft food, but even so, the jaw is wider than when seen on the surface and extends back behind the ears.

From the top, the skull is a warped oval shape. There are indentations making eye sockets, and the eyebrows jut out, casting shadows over the eyes. The cheekbone sticks out, creating a level area. Around the mouth and the chin area below, the face gently curves. The cheekbone extends to the ear, creating a small, hard protuberance along the way. One effective way to portray the elderly is to accentuate this protuberance.

The chin area on women is closer to the neck than it is on men, and it is often portrayed as jutting farther forward and higher than it actually does. By clearly portraying the jaw as large, the head will look more alive. Also, take care to show the front teeth jutting out farther than the jaw.

The slopes from the nose to the tip of the jaw and from the forehead to the end of the nose are important in drawing the face.

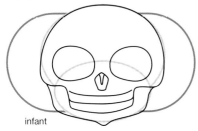

aging of the skull

infant

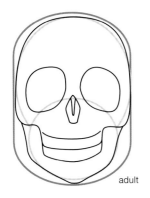

adult

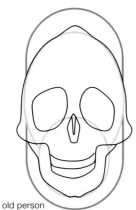

old person

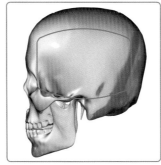

gender differences in the skull

male

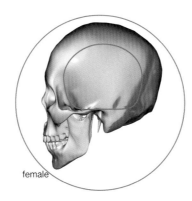

female

age/sex differences in skull geometry

First, let's look at gender differences. From the side, the male skull is more rectangular than the female one. The jaw is bigger and the forehead is flatter than the female skull. The female skull has a smaller jaw, and the chin is pulled back. It is also rounder; from above it looks more spherical.

Next, let's turn to age differences. Our brain and eyes are large from the time we are born. They don't show the remarkable growth of other body parts. A baby has big eyes and a big head, but the jaw is very small. It almost looks like it has been smashed in on the sides. As we age, the top of our head becomes more pointy, the teeth fall out, and the jaw becomes smaller as it degenerates. The head becomes longer and thinner.

head muscles ☛2

The head has muscles to help the face convey expressions and to help the jaw chew. For more on the expressive muscles, turn to the face chapter beginning on page 102. The chewing muscles are the temporalis, a large muscle that spreads across the side of the head; and the masseter, which extends from the cheekbone to the jawbone.

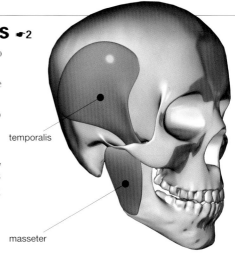

temporalis

masseter

☛2. Tension in the temporalis makes veins stand out

The temporalis covers most of the sides of the skull, goes inside the cheekbone, and attaches itself to the coronoid process on the jawbone. You can see this muscle in the temple when it is tense. When it does tense up and the blood pressure starts to rise, the veins around the temple can be seen on the surface.

The muscles used to open the mouth are in the neck. These muscles make the complex movements of the jaw possible.

h
head

Structure

Exploring the
Construction of
Human Body Parts

Part 1

head

placement of head parts

From the front, the eyes are along a horizontal line right in the middle of the head. The area just below the nose comes halfway between that line and the end of the chin. About halfway between the end of the nose and the chin is the lower lip.

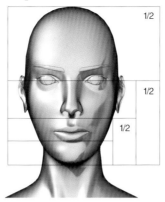

placement of ears ☞3

From the side, the ears begin on the vertical center line. The top of the ear just touches an imaginary horizontal line from the outside corner of the eyes.

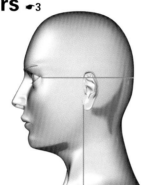

beards and mustaches ☞4

There are many different kinds of beards. Two key points to note: Does the beard continue from the ears to the chin? And, is the beard on the chin connected to the mustache on the upper lip?

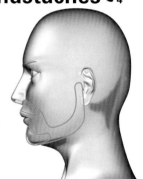

triangular arrangement of facial features

Draw a line from the middle of the mouth to the outer edge of the eyes, and you'll have an isosceles triangle. The longer the two equal sides are, the more adult the person is likely to be. The longer the base line, the more likely the face is that of a child.

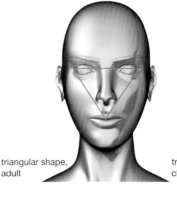

triangular shape, adult

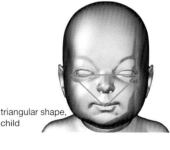

triangular shape, child

☞3. Exaggerations in *manga* and anime

Look at the "placement of ears" diagram to see just how far apart the eyes and ears are. From the side, the eyes are almost completely looking forward. Formerly, to stress the emotions or individuality of a face, Japanese cartoon and cell animation creators would draw the eyes bigger and point them more to the side, making it easy to see them in profile. Once you get used to that, a real profile—where the face is level and looking forward, and you can't really see the eyebrows, eyes, and mouth—might be a little unsettling.

The surface of the head

Hair grows on the head. It protects the head from collisions and sunlight and softens the differences in temperature. The part of the skin covered by hair hardly gets any sunlight. You can see this effect by looking at your hairline. The skin on the face may look smooth, but it has pores, shadows, and tiny bumps on it.

☞4. The surface of the face

On a man's face, the pores stand out where his beard grows, even if he is clean-shaven.

Women use cosmetics, which give their face a scoured, neat look; the makeup goes on the face almost like an artist puts paint on a canvas. There is a big difference in the kinds of paintings women present. For more on cosmetics, turn to the face chapter beginning on page 102.

jaw movements for opening/closing the mouth ☞5

The jaw is the biggest mobile part in the head. Because it slides forward when opening, the mouth can open over a long vertical line.

jaw movements in lateral directions ☞5

The jaw can also rotate from left to right. This allows the back teeth to grind food.

changes caused when opening the mouth ☞6

When the mouth opens, the head does not move. The muscles to open the mouth are in the neck. When the mouth opens, a node appears just before the ear and moves downward.

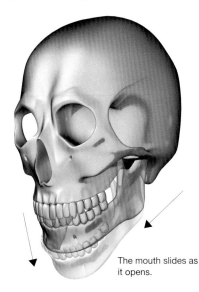

The mouth slides as it opens.

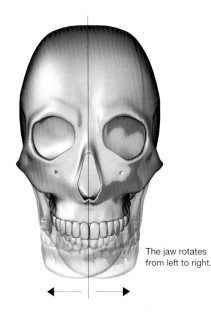

The jaw rotates from left to right.

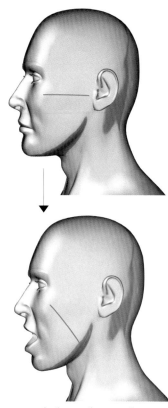

As the mouth moves, the node changes.

tilting the head forward/backward ☞7

When lifting your head, you lean forward and jut out your jaw to keep balance. When you hang your head, you round your shoulders and drop your jaw.

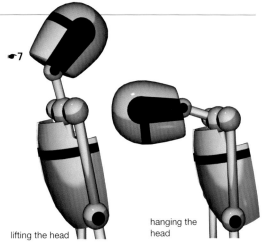

lifting the head

hanging the head

☞5 ☞6. Opening/closing the mouth

When the mouth opens, the head doesn't move, but the neck changes because the muscles used to open the mouth are located there. The jawbone is shaped so that when you open your mouth wide, a node appears just before the ear and proceeds to slide down. The jaw has a very strong bite; it often closes much faster than it opens.

Next, let's look at the up-and-down and side-to-side movement of the jaw. When you raise your head, your mouth opens, unless you consciously close it. This occurs because the muscles in the neck used to open the mouth are pulling at the jaw. Often, you will part your lips without even being conscious of it, but even if you keep them closed, the jaw will drop, making its outline slightly longer than normal.

☞7. Balancing when you lift or hang your head

When you raise your head, you usually lean forward and jut out your jaw to shift your center of gravity. If you thrust your head back with your back straight, you'll fall over (unless you're sitting in a chair). When you look down, you push your back out a little and pull in your jaw, and your neck looks like it is suspended from above. When you hang your head, you round your shoulders and drop your jaw. When you turn your neck to the left or right, the jaw is drawn in and the shoulders stay still. When you look back, the neck turns, the jaw juts out in the same direction, and the shoulders follow.

h
head

Structure

Exploring the
Construction of
Human Body Parts

Part 1

head

ranges of jaw movements

The jaw slides when it moves. From the side, the point where rotation begins is slightly behind the wisdom teeth. With that spot as the center, the jaw can move no more than 40 degrees, starting from a closed position.

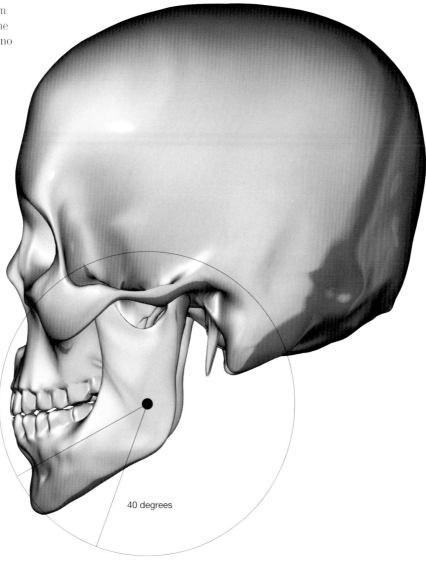

40 degrees

The important things to remember when depicting the movements of the head are whether you can accurately express a person's interests by frankly portraying the act of being attentive, and whether you can capture the contrast between the stiff movements of the head and the pliant movements of the neck supporting it.

☛7. Head movements

When the head moves, pay attention to movements in the waist and shoulders. It's unnatural if the shoulders don't even quiver when the head moves. The head, well above the body's median line, has all sorts of restrictions on it to maintain balance. Often, head movements are instantaneous and precise. Of

course, there are times when the head moves slowly, but this cannot take place continuously. The eyes, the biggest information collection center in the head, demonstrate their maximum ability when they are still. Also, the ears find stillness more desirable for hearing things with maximum clarity. Basically, if you are looking at something and not moving,

your head is in a fixed position. When something else moves into view, you quickly look toward it. Unless there is a special reason, the head moves before the rest of the body.

mouth

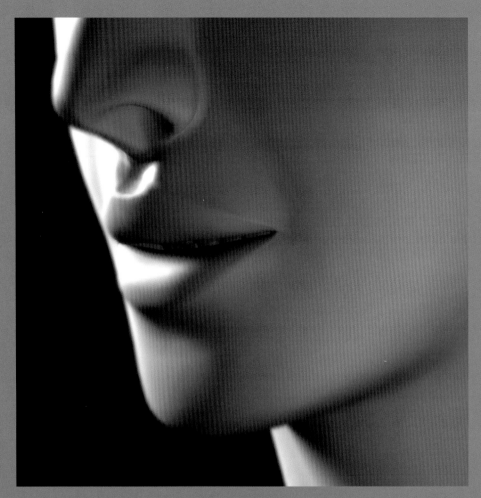

mouth

The mouth is the biggest of the seven holes on the face. The nose, ears, and eyes all have two holes each, but there's just one mouth. For the eyes and the ears, which have to take in information about their location in a three-dimensional way, having two of each is a necessity. The nose is important in breathing, so it may be that there are two nostrils just in case something happens to one that would stop it from being able to breathe. The mouth is called the entrance to the vertical tunnel of the body, as it leads down into the digestive system. It has three roles: talking, eating, and breathing. While the nose is usually responsible for breathing, when you need a large volume of oxygen all at once, you use the mouth.

Related pages   p. 90 nose & ears   p. 102 face   p. 170 looking back

m
mouth

Structure

Exploring the
Construction of
Human Body Parts

Part 1

mouth

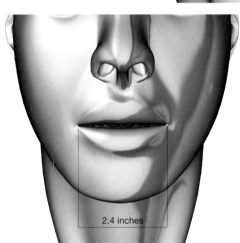

3.5
inches

1 inch

upper lip   lower lip

mouth shapes and proportions

Example: 25-year-old female
The jaw decides the outline
of the face. The rows of teeth
control the shape of the jaw.
The male jaw is more robust;
the female jaw is smaller.
Women sometimes have trouble
with their teeth coming in
properly because of their small
jaws. From the tip of the nose to
the chin is about 3.5 inches.
There is about 1 inch between
the top and bottom lips. The
mouth is about 2.4 inches wide.

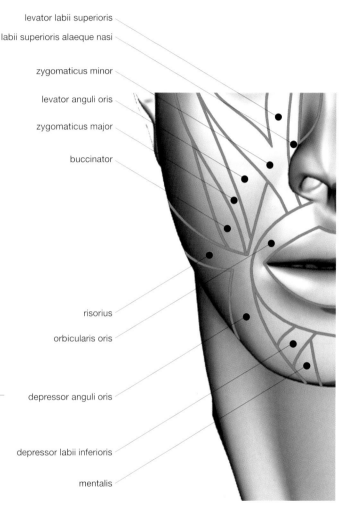

levator labii superioris

labii superioris alaeque nasi

zygomaticus minor

levator anguli oris

zygomaticus major

buccinator

risorius

orbicularis oris

depressor anguli oris

depressor labii inferioris

mentalis

2.4 inches

mouth muscles ☛1

The group of muscles that
make up the face's expressions
can change the mouth in minute
ways. Be sure to capture how an
expression transforms the face.

**☛1. The expressive muscles
around the mouth**

The temporalis and masseter
muscles are in the jaw. For more
detail, see page 80 of the "head"
chapter. Here we will look at the
expressive muscles around the
mouth. The face has an intricate web
of muscles on it, and the area around
the mouth is the most complex. These
muscles radiate out from the mouth's
edges: the orbicularis oris, levator
labii superioris, zygomaticus minor,

levator anguli oris, zygomaticus
major, buccinator, risorius, depressor
anguli oris, depressor labii inferioris,
and mentalis. Also, the orbicularis
oris encircles the mouth.

These muscles offer a hint as to
how an expression transforms. They
aren't the sorts of muscles that move
the body or create power; they use
just a little strength to move the skin
on the surface and change the way
you look.

**☛1. Movement of the
expressive muscles**

These muscles move in a very
complicated way. First, the jaw and
the expressive muscles usually move
in sync—when we talk and eat,
especially. Try this simple test: Look
in the mirror and, with your mouth
closed, try to move your jaw without
moving any of the expressive
muscles. You probably don't look like
you're eating. If you do the
opposite—exaggerate the movement

of those muscles while you open and
close your mouth—it looks more like
you are eating. For example, when
you chew something, you close your
mouth, pull your lips in, and slightly
narrow them. There's no space
between your front teeth and your
lips. The cheek takes the same sort of
action. The movement is over in a
second, so it's difficult to pick up, but
it's good to keep that in your mind.
Eventually, you'll naturally be able to
see different movements as well.

85

inside the mouth ☛2

The expression of the mouth is decided by the rows of teeth inside and the jaw. The tongue can transform freely. When you are trying to depict it, think of *tarako* (salted cod roe).

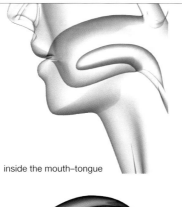

inside the mouth–tongue

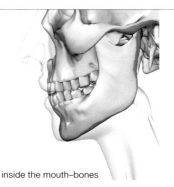

inside the mouth–bones

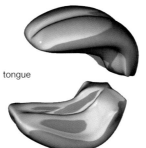

tongue

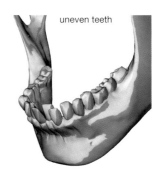

uneven teeth

teeth ☛3

There are 16 teeth on both the top and bottom (children have 10 on each level). The thin, sharp front teeth bite off food; the wide, flat molars grind down.

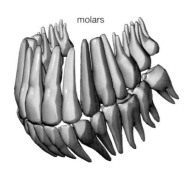

molars

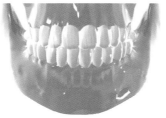

surface of the teeth

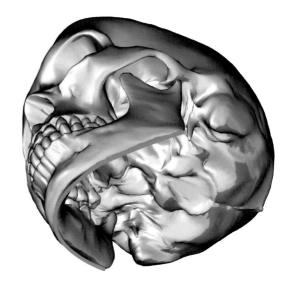

☛2. Tongue and lips

From the side, it looks like the top lip sticks out further than the bottom one. The line along the bottom of the lower lip gently sinks. From the front, the indentation just below the nose looks like a V; the bottom of the V looks slightly swollen. The bottom lip lies gracefully; the part just below the middle of the lip looks as if it has been raised. The tongue is big and has tendons running lengthwise in the middle.

The inside of the throat has folds on both sides along the top. The uvula is in the middle. In the far back at the top are the passages for the nose. Below is the path to the esophagus. The tongue can fold itself back and become like a lid on the throat, defending it from objects in the mouth.

☛3. Special characteristics of the teeth

Teeth have very long roots that are deeply imbedded in the gums. There are 16 teeth on the top and bottom for a total of 32. Children have smaller jaws; they have 10 teeth on the top and bottom for a total of 20.

The front teeth are thin and sharp to bite off food; the back teeth are flat and wide to grind down food; and the canines are shaped like fangs for grasping and holding on to food. The jawbone is thick and extends to just below the ear, where it bifurcates. Remember that the space below the jawbone is open. There is no bone; it is soft, and fat easily accumulates here. Also the movements of the tongue and cheek are detailed here.

The teeth have the dull, transparent sheen of enamel on them. The surface has little lengthwise wrinkles on it, giving the teeth a complex luster.

m
mouth

Structure

Exploring the
Construction of
Human Body Parts

Part 1

mouth

lips and skin around the mouth ☛4

When depicting the face, you can't forget the contrasting wet and dry places. The differing temperament of the skin and areas around the lips is important to capture.

mouth surfaces ☛5

There is a glut of tiny capillaries on the lips. The skin is thin, and the surface is sensitive. The inside of the lip and the mouth is wet with saliva; the surface of the tongue is gritty.

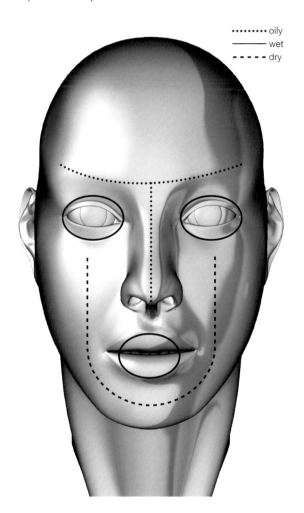

········ oily
———— wet
- - - - dry

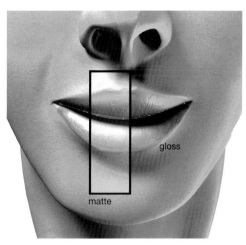

differing temperament of the lips

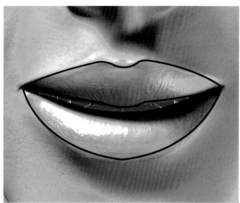

lip line

☛4. A key point: the wet and dry places

The inside of the lip and the mouth is wet with saliva; the surface of the tongue is gritty. On the back of the tongue, there are two bumps and a sinew.

The only two spots on the face that are consistently wet are the eyes and the inside of the mouth. They are clearly of different substance than the other areas of the face.
The area from the forehead to the bridge of the nose is oily. It is sometimes shiny because of the oil secreted by the sweat glands. This is different from the clearly wet surfaces of the eyes and mouth. The fact that these places are wet means they can't be in constant contact with air. That's why the eyes blink and the mouth is usually closed.

The lips are easily chapped. Women usually put on lipstick or balm. Men often lick their lips when they are dry because it feels uncomfortable. The difference between the lips and the skin surrounding the lips is important to note in men, too.

☛ 5. Makeup and the lines of the lips

The outline of a woman's lips is complicated to depict. Women usually wear makeup, so there are many possibilities to consider: Do you create the two mountains on the top lip? Should the lips be full or thin? Will the lips be highlighted or more natural? In other words, the way you decide to outline the lips also decides their shape and the mood they create. And don't forget to decide whether the lipstick will be matte or gloss.

mouth movements ☞6

Generally speaking, there are two sorts of mouth movements: the kind when the jawbone moves, transforming the whole face; and the kind where the expressive muscles make subtle changes in the skin around the mouth.

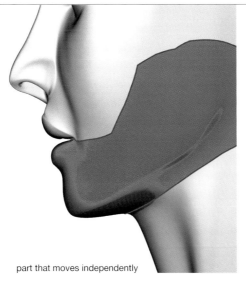

part that moves independently

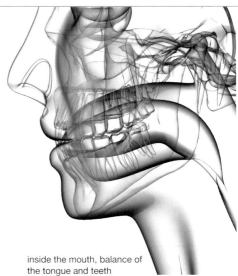

inside the mouth, balance of the tongue and teeth

mouth shapes and characterization ☞7

A yawn indicates boredom. A slack-jawed look makes people seem stupid. If the mouth is slightly open, a person seems inattentive. A closed mouth gives the impression of wisdom and a strong will.

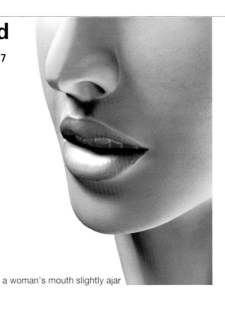

a woman's mouth slightly ajar

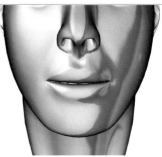

closed mouth

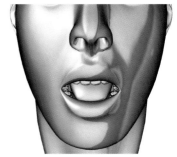

opened mouth

☞6. Movement of the bones in the jaw

Humans are basically omnivorous. If our teeth can cut the food, we can digest it. That's the way we are made. The front teeth cut the food; the back teeth grind it; and saliva dissolves it. The jaw moves up and down when using the front teeth; it moves from side to side when chewing with the back ones. Often, the jaw moves independently from the neck or head.

☞7. Is an inattentive woman's mouth open?

People yawn when they lack oxygen and want to replenish their supply with one big intake, but we still feel that a yawn is a sign of boredom. The yawn shows the range of movement the mouth is capable of—the average one lasts a lengthy six seconds.

Strangely, if we leave our mouths open, we give the impression of not being smart. A closed mouth conjures up images of a strong will and intelligence. Normally, when we are not paying attention, our lips are closed but our jaw is slightly open. The tongue spreads out over the inside of the mouth, touching the back of the teeth. If we open our mouths ever so slightly, we appear approachable. This look is often given to a beautiful woman to make her seem friendly or to add to her

sexuality. The mouth has a lot of sex appeal, and some people try to cover it at times because they are bashful about having it seen.

The pitfalls of lip-syncing

Charts showing the different mouth shapes for pronouncing vowels or special sounds represent general standards. While they are important to learn, they don't really show the actual movement of the mouth. If you

m
mouth

Structure

Exploring the
Construction of
Human Body Parts

Part 1

mouth

tooth diagrams

Both baby and permanent teeth form curved lines on the top and bottom of the mouth. The permanent set consists of 32 teeth, as shown in the diagrams below. The teeth are symmetrically aligned from top to bottom and left to right.

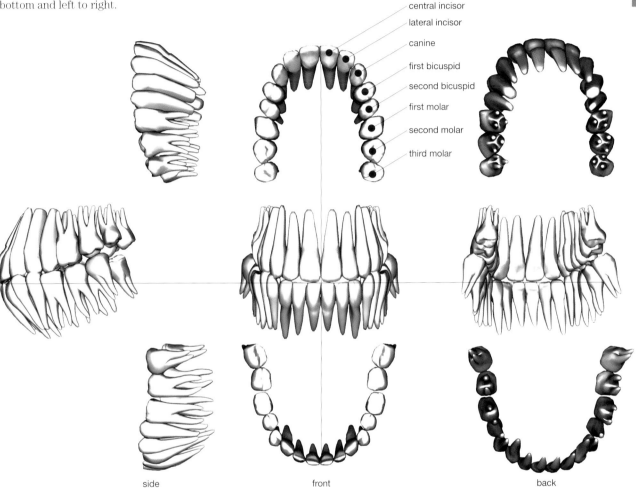

central incisor
lateral incisor
canine
first bicuspid
second bicuspid
first molar
second molar
third molar

side   front   back

begin to think that the mouth really makes those shapes when saying something, you may end up making animation where people just string those mouth shapes together as they talk, like a child lining up toy blocks.

One of the bylaws of animation when drawing human beings is that to create something natural, it is better to concentrate on reading the expressions that convey someone's feelings than it is to focus on each

syllable. The single letter "a" can be pronounced in an unlimited variety of ways; it's not necessary to conscientiously depict each sound. Take sounds that burst forth like "ba" or "pa," where the mouth moves the most. The only mouth shapes it is necessary to depict are when the lips come together for just a second and when the jaw opens.

Pronunciation should take a backseat to laughter, surprise, and

other emotions when it comes to drawing mouth shapes. If a character turns to the person next to him and says, nonchalantly, "Do you want to get going?" the lips move in a vague way. If you depicted the mouth saying each syllable as it is supposed to be pronounced, it would be a very frightening sentence. "Lip-sync" may be better characterized as "emotion-sync" because it is closely tied to expressions and feelings.

nose & ears

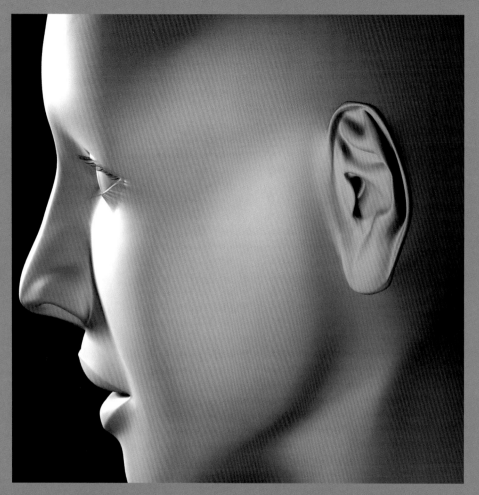

The nose and ears send signals to the brain as they provide two of the vital five senses. Because there are two ears, the brain can process sounds three-dimensionally. We would lose an incredible amount of information in a world without sound. It's like watching the TV without the sound on. With the nose, we can smell things from a distance and immediately infer when danger is near, for example. Even more importantly, the nose helps us breathe and regulates the temperature of the air coming into the body. The nose is an important organ, but it moves little and is rarely studied (it's liable to be omitted or abbreviated in pictures).

Related pages   p. 170 looking back

n&e
nose & ears

Structure

Exploring the
Construction of
Human Body Parts

Part 1

nose &
ears

nose/ear proportions

Example: 25-year-old female
The nose is about 3.15 inches long and projects some 1.18 inches from the face. The wings of the nose are about .79 inch tall. The ears are 2.76 inches long and 1.38 inches wide. The section of the ear connected to the head is about 2.36 inches long and 1 inch wide.

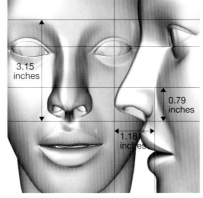
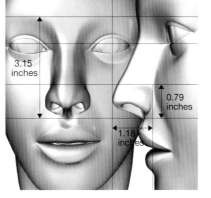

nose/ear geometry ☞1

The parts of the ear offer guideposts to understanding its boundaries. The shape and size vary greatly depending on racial and individual differences.

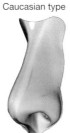
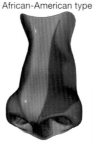

Caucasian type   Asian type   African-American type

types of noses by race

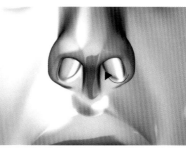

curling in from the wings of the nose

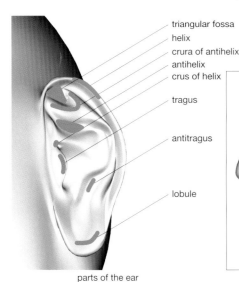

triangular fossa
helix
crura of antihelix
antihelix
crus of helix

tragus

antitragus

lobule

parts of the ear

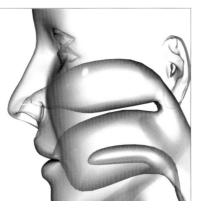
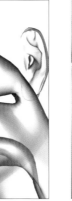

nasal cavity

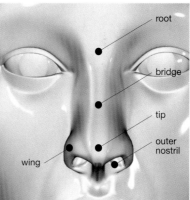

root

bridge

tip

outer nostril

wing

parts of the nose

☞1. The shapes of the nose and nostrils

The nose is in the center of the face, and it is the face's most prominent protuberance. It also casts a big shadow on the face; the nose abounds with different expressions depending on how the light hits it. The height of the nose is partially determined by race. Caucasians tend to have tall, thin noses; African-Americans have flat, wide noses.

The higher the root of the nose, the more the bridge seems sharp—a sign of beauty. If the root is low, the face seems flat. The end of the bone is in the nose bridge; you can see a subtle change when it turns to cartilage. The tip of the nose has a slight indentation at the very end. The wings puff out on either side and their size can be altered. Inside the nostril, the backs of the wings flare out a bit. Notice that the line that begins at the end of the wing and curls back in continues up and toward the center of the nose. The shape of the indented area between the middle of the nose and the mouth is also complicated. Remember to check the profile to see the height differences. The nostrils quickly connect with the nasal cavity. That's why they look dark inside.

☞1. Knowing the boundaries of the ear

The ear is a complex shape and subject to individual differences, making its characteristics hard to grasp even after studying it. Inside the ear, the eardrum captures the vibrations from sounds. Deeper inside is an organ that controls the sense of balance. The ear we see on the outside is the umbrella that gathers sounds. The names of the ear parts are in the diagram above. While the names aren't that important for our purposes, the parts themselves are the guideposts to the shape of the ear. The lobule, known more commonly as the earlobe, completes the ear's outline; it varies greatly depending on the individual.

nose/ear cartilages ☛2

The bone in the nose only extends part of the way. The rest is cartilage. That makes it soft, allowing the nose to take on many shapes reflecting a person's mood.

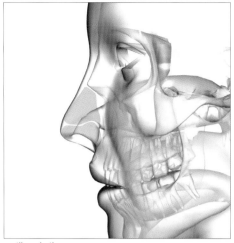

cartilage in the nose

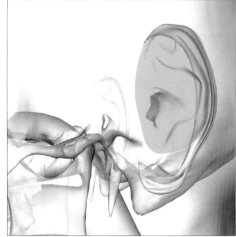

cartilage in the ear

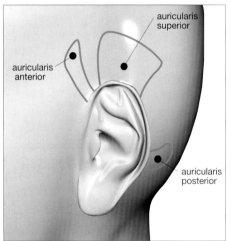

auricularis superior

auricularis anterior

auricularis posterior

ear muscles

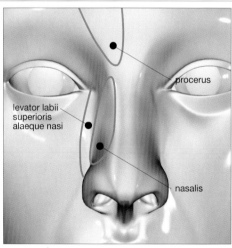

procerus

levator labii superioris alaeque nasi

nasalis

nose muscles

nose/ear muscles ☛3

The nose has muscles that create expressions such as the nasalis, on both sides and toward the bottom, and the procerus, between the eyebrows. Also, a muscle that greatly changes the nose's shape is the levator labii superioris alaeque nasi, on both sides and running along the upper part of the nose. These muscles do more than change expressions by creating wrinkle after wrinkle; they also allow the nostrils to open and close.

There are practically no muscles that move the ears.

☛2. Cartilage creates the shape of the nose and ears

The skin on the nose is thin, and the cartilage inside is almost the same shape as the nose. In a nose with a well-developed cartilage, a ridge develops and it looks as if the nose has two levels. The cartilage is soft and can change the nose's expression as well as its shape, but the part with the bone cannot move. The skin around the nose pushes up against it and sometimes creates bumps. Typically, once an expression changes, the skin on the surface of the nose becomes hard and adds liberally to the expression.

The ear has no bone and thin skin. The cartilage creates the ear's shape. There is no cartilage in the earlobe.

☛3. Only a few people can move their ears

On both sides of the head at the base of the ears, the auricularis anterior, superior, and posterior muscles are found. People who can manipulate these muscles and move their ears are a rare breed. More typically, the ears move because the mouth has been opened wide and closed. In other words, it's the influence of the jawbone. The earlobes are influenced the most; the top part doesn't move much. For those few who can move their ears voluntarily, the upper part of the ear moves the most.

n&e

nose & ears

Structure

Exploring the
Construction of
Human Body Parts

Part 1

nose &
ears

nose/ear surfaces

The nose, especially the tip, is an oily place that sometimes has a shine to it. The surface of the ear is covered with lots of tiny hairs. The skin is thin, and blood vessels can be seen.

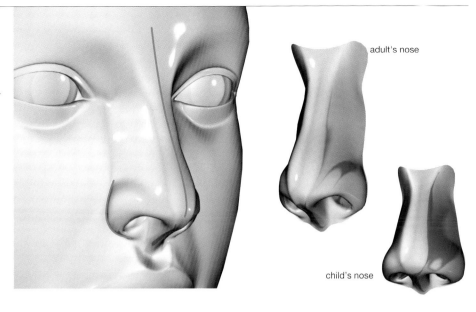

adult's nose

child's nose

wearing eyeglasses ◄4

To create a person wearing glasses, you have to precisely understand the placement of the ears and nose.

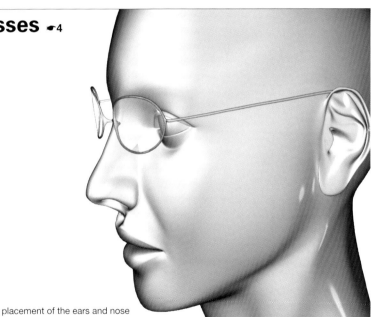

placement of the ears and nose

◄4. Glasses tell us about the placement of the ears and nose

If you aren't clear about the whereabouts of the nose and ears, you won't be able to put on your glasses. The glasses sit on the bridge of the nose, and the handles extend to the ears. The handles are usually horizontal and parallel, and while the length is up to individual preference, handles don't vary that much.

nose/ear movements 5

The ears can't move on their own. They move when the head tilts to hear something. The nose doesn't seem like it moves, but it is teeming with expressions.

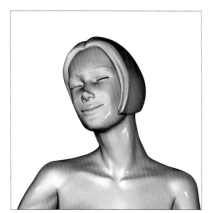
smelling and expression

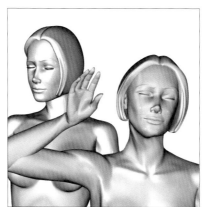
different ways of hearing

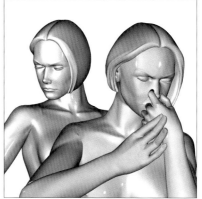
rejecting a smell

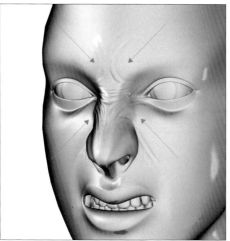
nose wrinkles

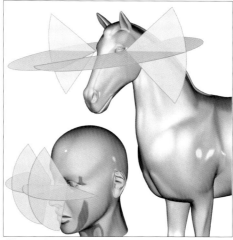
difference between humans and animals in range of senses

5. Ear movements are head movements

Consider the act of straining one's ears. The head lifts slightly when trying to take in the surroundings or hear a sound. If we are trying hard to pick out sounds, we lower our head. If we are intent on hearing a faint sound coming from one direction, we tilt an ear in the direction of the sound, trying to bring the ear closer to the source. On top of that, when people put their hands near their ears, they are usually not trying to

hear better; they are trying to hear anything at all, or they are making a sign that they want to hear something repeated. An invisible action like hearing sometimes calls for a dramatization, like using the hand to exaggerate the process.

There is one more hearing action that can't be overlooked: losing one's focus. In other words, usually when we are trying to hear something, we don't see anything. As we listen, our line of vision turns to the sky. We sometimes subconsciously let our

line of vision float here and there, but basically, when we are listening to something, we stop looking.

Separating the sounds we want to hear from the sounds we don't

When we hear a sudden burst of sound, we quickly turn toward the source. The ear is usually open to sounds from all directions; it separates the sounds we want to hear from the other sounds we can hear. Unless a very loud sound

continues to drown out everything else, people don't press their hands against their ears. But if we are enduring a loud sound, our mannerisms show it right away. The shoulders flinch; the face grimaces; the head moves to the side and down.

Objectively, it is hard to tell whether a person is listening or not. But when people try to listen to something, they commonly turn from their previous line of vision and stay quiet and still.

n&e

nose & ears

Structure

Exploring the
Construction of
Human Body Parts

Part 1

nose &
ears

nose shapes and expressions ☞6

The nose changes shapes and adds lots of
wrinkles on all sorts of occasions—when we're
angry, laughing, sad, sneezing, yawning, and
sniffling, for starters.

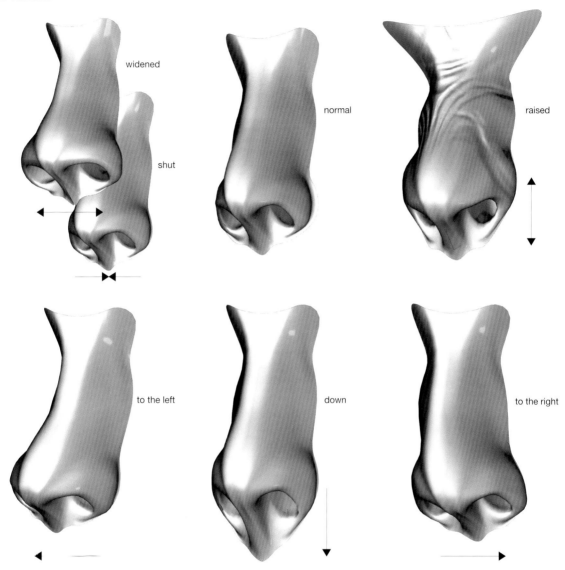

widened

shut

normal

raised

to the left

down

to the right

☞6. Depicting the emotions and movements of a wrinkly nose

The wrinkles on the top of the nose of a frightened person are mostly horizontal and between the eyes. The skin pulled from the cheeks makes vertical lines on both sides of the bridge of the nose. The skin on both sides of the nose is raised, and the wings of the nose also move up quite a bit. One note of caution: The cheek actually changes more than the nose in this case. When we are sad or worried, the wrinkles gather between the brows.

The one area of the nose that moves the least is the tip. When we smell something, our head barely moves. Just as when we are listening, we first lose our line of vision, and then we forcibly breathe in through our nose little by little. We also see the cartoonish version where people jut out their jaw and lift their nose up to smell something, but people don't actually do this very often. At any rate, the act of smelling something is often accompanied by a change in expression. We smile when we smell good food; if we're suddenly hit with a smell from our past, we may get sentimental and even shed a tear. The sense of smell is difficult to sum up in words, and we're liable to recollect all sorts of memories through it.

☞6. Passive organs?

While vision is an active thing, smelling and hearing are more passive. But in the early years of human evolution, those senses must have been active too. Smelling food allowed us to know if it was rotten or not, and whether we could eat it. And the ears would have allowed someone to quickly detect a natural enemy in the area. Thanks to progress and evolving societies, the roles of smelling and hearing are different today. The nose's biggest function is to assist the sense of taste; the ears are used mainly to listen to music and to take in verbal information.

skin & hair

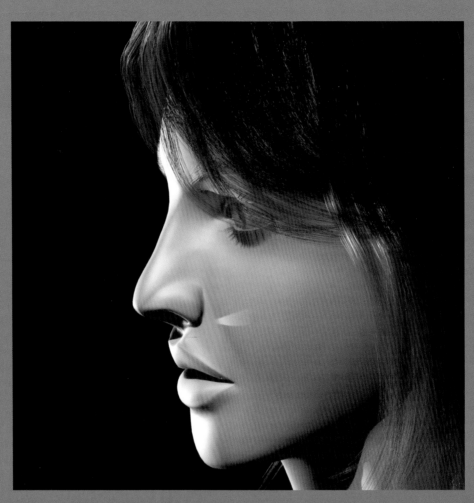

The skin breathes as if it had a life of its own. It covers the surface of the body and expels grime, sweat, and other unneeded substances. It protects the body from impact, regulates the body temperature, and also feels fever, pain, and other sensations through the sense of touch. Skin is also capable of self-purification and self-repair. Like a parasite, it helps the host. Skin is alive, but hair is dead. Hair, like human nails, is made of dead cells. Hair protects the body and its temperature, and aids the sense of touch. Hair, which stops sweat at the surface for a long time, works in partnership with the skin to maximize their protective qualities. In this chapter, we'll explore the structure, function, and connection to body movements of the skin and hair.

skin & hair

s&h
skin & hair

Structure

Exploring the
Construction of
Human Body Parts

Part 1

skin &
hair

skin structure ☞1

The epidermis is at the top. Below it is a spongy layer of dermis. Below that is the fatty hypodermis. The outer layer of the skin, the stratum corneum, regenerates every month.

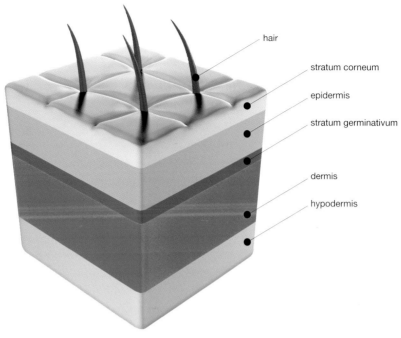

hair

stratum corneum

epidermis

stratum germinativum

dermis

hypodermis

sex differences in skin conditions

Women have more hypodermis, giving their bodies beautiful curves. For men, the muscles stand out more.

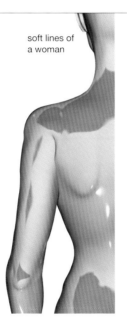

soft lines of a woman

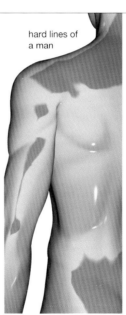

hard lines of a man

☞1. Functions of the different layers

Skin consists of three layers. The outer layer is the taut epidermis. Below that is the spongy dermis. And below that is the fatty hypodermis. On the outside of the epidermis is the stratum corneum. This layer is born from the thin layer between the epidermis and dermis—the stratum germinativum—and regenerates on a monthly cycle. The stratum corneum

is dry. Oily membranes secreted by the sebaceous glands can be found on the surface; they help keep the moisture in the skin from evaporating.

The dermis protects the surface, and its cells help make up the stratum corneum. The dead stratum corneum cells turn to grime and are discharged from the body. Capillaries run through the spongy dermis, giving nutrients to the cells. The thick vein along the main route of the capillaries

can be seen on the surface.

The hyopodermis is very well insulated; it does its best to keep the body's temperature stable. It stores excess nutrition and guards against hunger. It also is effective in absorbing impact and responsible for creating the elegant lines on our bodies. Of course, if too much nutrition is stored, the results are anything but beautiful.

Skin and hair can't move on their own

Skin can't move of its own volition. Also, hair is moved by the body's movement or by outside forces like the wind. When drawing the head on the hair, for example, you need to relay its weight and volume to the viewer.

hair structure

Hair ends are fine. They draw nutrients from the blood vessels at the root of the hair. The surface is full of oily membranes, and there are sebaceous glands at the roots. Melanin, produced by pigment cells in the epidermis, gives hair its color.

hair

skin

sebaceous gland

blood vessel

hair surfaces

Magnify a strand of hair and you'll find scaly bumps heading out toward the tip. This makes it easy for sweat and oil to glide down the hair. The texture and feel of hair differs greatly from person to person, as well as depending on what part of the body it is found.

straight hair

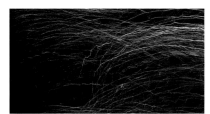
frizzy hair

wavy hair

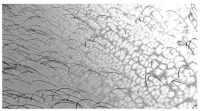
short hair

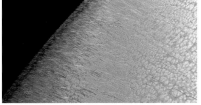
downy hair

Hair comes in bundles of five

The small amount of oily membranes on the surface prevents tangling and helps repel water. Hair grows in bundles of about five. Each person has about 100,000 strands of hair, and it goes without saying that most of these strands are crowded together on the head. Hair is quite strong; one strand can support an object weighing about 1.8 ounces.

There are places with no hair, such as the palms of the hands and the soles of the feet. When sensory ability is the priority, the hair does not have any melanin pigment.

Hair grows all over the head

We have so much hair that even though one strand is very light, hair collectively has an incredible weight. It tries to maintain its shape. Once it becomes long, it is often messy and split ends or gray hairs can be seen.

People have many different sorts of hair: straight, wavy, and frizzy, for starters. When making a model, remember that hair doesn't grow only on the top of the head—it grows all over the head. Small hairs grow on the forehead and elsewhere. The hair on the top of the head tends to be finer; lower down, the hair is thicker. Whorls, or bits of hair that have curled together, can often be seen on the head; they're usually a little bit behind the top of the head, but the number of whorls and their placement differ from person to

person. The skin at the top of the head rarely sees the sun.

100 people, 100 hairstyles and colors

If we do nothing with our hair, it will fall straight down or stick out in all sorts of ways. Usually, hairstyles are formed intentionally; they are often governed by the latest fashion. On the surface of the hair, a halo-like effect can sometimes be seen as the light creates a circle there. Because of the

s&h

skin & hair

Structure

Exploring the
Construction of
Human Body Parts

Part 1

skin &
hair

head hair surfaces ☞2

Head hair is covered by scaly cuticles. If the condition is bad, the hair creates split ends at the tip or the strand breaks off in the middle. The part of the surface with the oily membranes that shines is called "an angel's halo."

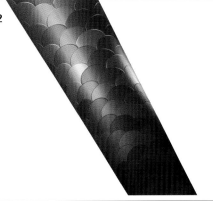

eyebrows ☞3

Eyebrows are not quite .4 inch long. One eyebrow has about 700 hairs. A lot of people alter their eyebrows, especially women, who lightly shave them and change the shape with an eyebrow pencil. The skin below the eyebrow moves a lot because it is closely tied to the face's expressions. Moving the eyebrows can exaggerate the expression in the eyes. The eyebrow follows the path set by the fine hairs in it. The hairs from the bottom and top meet in the middle and push out.

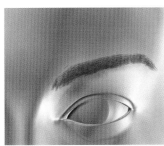

eyebrows

eyelashes ☞4

Eyelashes are about .5 inch long. One eye has about 170 lashes. About 70 grow on the bottom eyelid. They curve outward to prevent them from entering the eye and are quite strong.

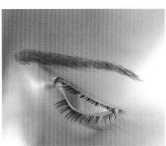

eyelashes

beards and mustaches ☞5

Beards and mustaches grow in many ways. Some people only have facial hair under their nose; others have it also on their jaw; still others have cheap-looking beards that make them look like a criminal. There are all sorts.

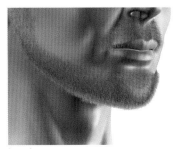

beard

nose/ear hair ☞6

Nose and ear hair is not often seen from the outside. Like eyelashes, these hairs keep dust and grime from entering the body.

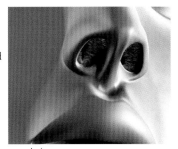

nose hair

oily membranes on the surface, the hair can show a radiance that is full of expression when it moves.

Hair changes color through dying or for natural reasons. Hair can be dyed all sorts of colors. It can also be manicured, or treated with slightly colored gloss. This gives the hair a tinted sheen when it's in the light. White hair is semitransparent, and it looks silver when it shines. It looks like a nylon fishing line. Don't try to express head hair in one color. Just

like hair length, hair color can vary widely.

☞2 ☞3 ☞4 ☞5 ☞6. The many roles of hair

The shape and role of hair depends on the part of the body it occupies. Some hair protects the head from impact. The cycle for creating hair lasts a long time, which is why the hair continues to grow out. Hair also insulates the body. Armpit and pubic hair is thick and curly, making it more

effective in absorbing impact. It is said that the more important the body part, the more likely it will be protected by thick, curly hair. Nose and ear hair keeps out the dust and grime in the atmosphere. The eyelashes take action before dirt or hair gets in the eyes. The eyebrows keep sweat from the forehead out of the eyes; they also shade the eyes, improving vision. And they play a big role in facial expressions.

Skin and hair draw people's

attention. They can be embarrassing or a source of pride. While some people may wear beards proudly, others may shave their underarms or legs to get rid of hair. With skin as well, some people like to attract attention by exposing flesh, while others may try to conceal themselves in their clothes, revealing little.

troubled skin ☜7

Skin is very delicate. Troubles such as tender or rough skin differ among individuals, genders, and age groups. There are too many skin blemishes to relate here, but they include acne, age spots, dull complexion, moles, clogged pores, and beauty marks.

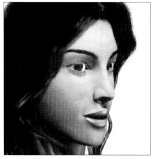

rough skin

aged skin

Aged skin is especially susceptible to wrinkles and age spots, as well as bumps created by clogged pores and color changes. The surface of the skin becomes just like the skin of an orange.

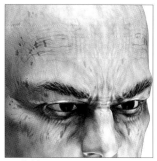

aged skin

close look at skin ☜8

The skin's surface is made up of little triangular or pentagonal ridges or bumps that surround the skin. This is the grain of the skin. When the grain is neatly aligned, the luster of the surface is more unified.

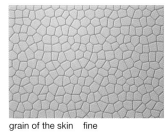

grain of the skin   fine

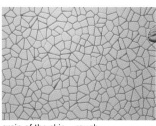

grain of the skin   rough

surface changes on the shin

suntan diagram

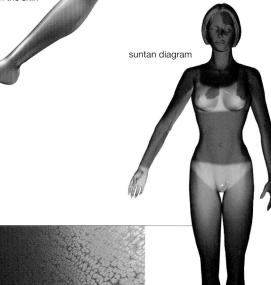

skin colors

The skin is slightly transparent, so the colors below the skin and the blood vessels can be seen. This gives the skin more expressive power and fills it with a color that is deep and seems almost self-luminescent.

skin changing color   white

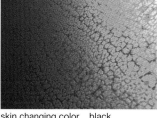

skin changing color   black

The skin turns dark because it releases more melanin pigment to protect the skin cells from harmful ultraviolet rays. That's what happens during a suntan. The lines between the exposed skin and the rest of the body are very clear.

☜7. The source of rough skin

Skin can easily turn rough on the face, the middle of the chest, the shoulders, the back of the neck to the shoulder blade, the palms, the shins, and the bottom of the feet. Rough skin can come from many sources, including spending a long time in the sun, stress, diet, shaving off extra hair, and hardening in the open air.

Color changes don't just happen because of time spent in the sun. Taking scrupulous care of your face may result in making your face slightly lighter than other exposed parts. With women, the neck is especially likely to be a different color than the face.

☜8. Depicting changes on the skin surface

The skin surface has minute wrinkles on it—this is the grain of the skin. Healthy skin is said to have a fine grain, but it isn't actually fine or narrow; it is more orderly and uniform in appearance. The skin has oily membranes that allow it to reflect sunlight, but the grain diffuses the reflection, giving the skin a complex, soft look. Women have a finer grain to their skin. Men have a rougher, more disorderly, and intricate grain.

Depicting the skin is difficult. Make it too shiny and it will appear hard; too many wrinkles will make it look dry and reptilian; bring out the veins and the colors beneath the skin and you risk creating something grotesque. Naturally, the fine hairs keep the boundaries from being clear.

s&h

skin & hair

Structure

Exploring the
Construction of
Human Body Parts

Part 1

skin &
hair

veiny skin areas ☛9

With white skin, the epidermis is somewhat transparent, and the veins and colors below come through to the surface.

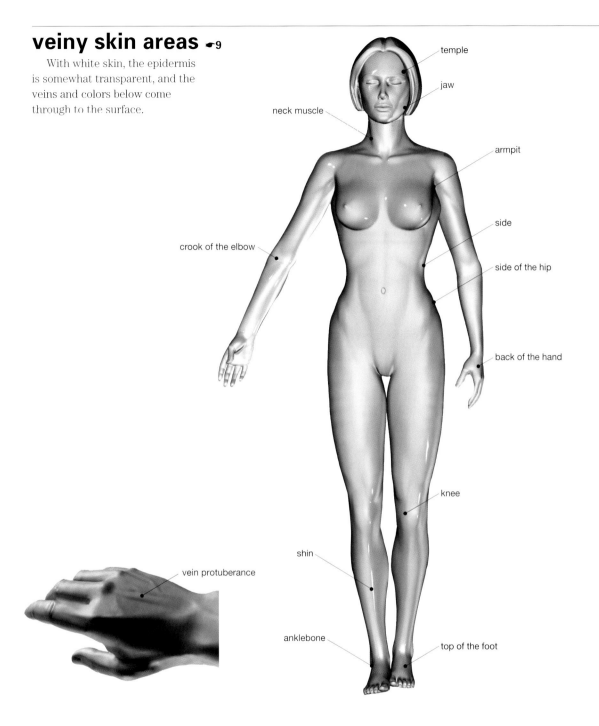

temple

jaw

neck muscle

armpit

side

crook of the elbow

side of the hip

back of the hand

knee

shin

vein protuberance

anklebone

top of the foot

☛9. Vein protuberances

Blood can collect in the veins along the forearms, hands, or tops of the feet, bulging against the surface of the skin. The face tends to be lighter because we take care of it every day; it is a noticeably different color than the neck. We don't fuss much with our neck and ears, so the ears' downy hair and the wrinkles on the neck appear as they are. Veins can be clearly seen in places where the skin is thin like the forehead and around the eyes, the neck, the backs of the hands and tops of the feet, both sides of the hips, and the insides of certain joints.

Swelling in the bath

When we take a bath, our fingers swell. Actually the rest of the body also swells. The fingers get waves of flesh on them because the nails stay unchanged and the extra skin gathers on the other side. The skin is very soft and can adapt to all sorts of conditions.

face

face

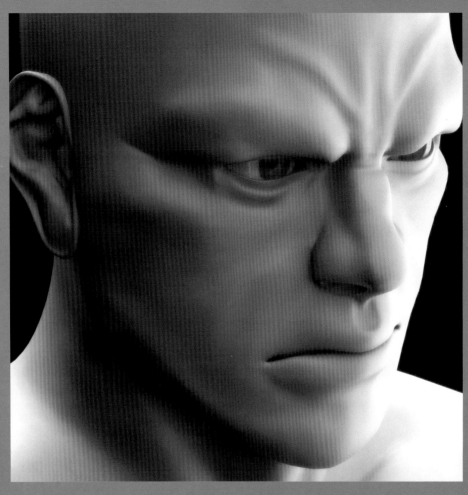

The face distinguishes individuals. Age and experience are etched in wrinkles and bones degenerate, but the personality expressed in the face remains. Facial expressions communicate various information, change in many ways, and are virtually impossible to classify into patterns. Changes are reflected in the character, emotion, environment, and present circumstances. Facial expressions are either encoded (intentionally shown) or unconscious (unintentionally shown). The former resemble a mask that can only be confirmed through a mirror, while the latter are instinctive and natural and can be seen within the former. This chapter explains the facial shapes, movements, and mechanisms.

grasping the facial shape ◄1

In creating facial expression, it is important to understand facial shapes. If you quadrisect the head, the face covers the lower front part. The cheeks are included with the eyes, nose, and mouth in an inverted triangle on the front. Grasping the facial shape from the cheekbone to the chin and the curvature from the middle of the forehead to the temple can be done by substituting them with simpler planes. One should pay less attention to the eyes, the nose, and one's personal sense of beauty. Rather, extra care should be given to the difference between the slope down the forehead and the nose by connecting the inclination from the mouth to the chin with a curved line A depression extends from the nostrils to both sides of the mouth. Also, the brow extends forward to create a shadow over the eyes. It is best to use a middle-aged person's face to easily understand the skeletal structure. The values below are based on a 25-year-old woman.

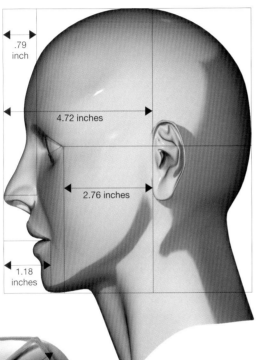

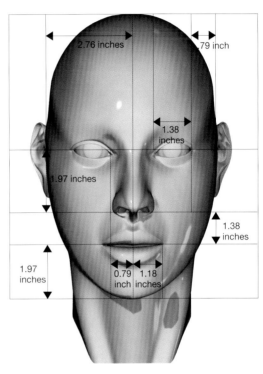

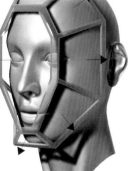

simplifying the planes

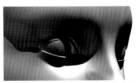

Shadows highlight the bulge of the brow bone.

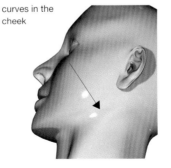

curves in the cheek

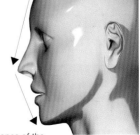

slopes of the face

◄1. Standard shapes are not appealing

Standard facial and body shapes have no special appeal. The wrinkles in the eyes' corners express depth of experience and geniality. A penetrating look communicates a firm presence and the force of will. Flaws add flavor to a person's uniqueness. The face reveals a person's age, gender, race, character, and emotions, which become the ultimate means of expression.

bulges on the skull ☜2

The eye sockets in the skull are hollowed out. They are big enough to range from the lower brow to the cheek. Also, protuberances can be found along the plane of the nose and cheekbones.

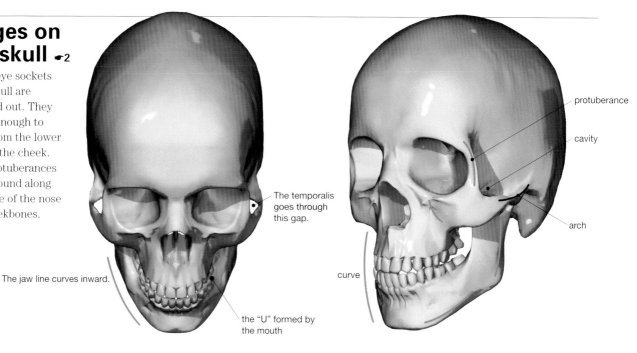

The temporalis goes through this gap.

The jaw line curves inward.

the "U" formed by the mouth

protuberance

cavity

arch

curve

facial muscles ☜3

The main muscles of the face are in the nose, mouth, and eyes. If you focus on muscles that change expressions, the busiest are the ones that move the brow and the upper lip.

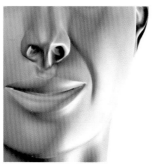

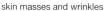

skin masses and wrinkles

procerus

levator labii superioris alaeque nasi

zygomaticus major

risorius

depressor labii inferioris

frontalis

corrugator supercilii

nasalis

orbicularis oculi

levator labii superioris

zygomaticus minor

orbicularis oris

depressor anguli oris

☜2. Protuberances on the head caused by bones

The hollowed-out eye sockets in the skull go from the lower brow to the cheek. This part runs up against the dent at the corner of the eyes. You can feel the curving bone between the corner of the eye and the brow. There is no bone at the temple. The flat area below the eyes gently blends with the protuberance at the nose. The flat surface that joins with the cheekbone juts out at an acute angle on both sides, then extends in a complex form toward the ear. The frontalis muscle is settled inside here. The dent below the cheekbones connects to the gums. The complex lines of the nose connect with the cartilage, and just below the nose is a big protuberance that connects with the teeth. From the front, both sides of the jaw follow the lines of the head and turn inward.

The line of the jaw heading for the chin tends to go in rather than up. The bottom of the jaw has no bones; the bones on the top of the mouth are complex.

☜3. The procerus and corrugator supercilii muscles between the eyes and the frontalis muscle in the forehead affect the eyebrows.

When the eyebrows move, they exaggerate the expressions of the eyes, although the effect is temporary. The frontalis covers much of the forehead; it provides the eyebrows' vast range and mobility. They move in a delicate but swift way. Rarely do eyebrows move slowly.

These same muscles help the upper lip rise, a common action that is seen most readily in a laugh. Both lips are pulled up and aslant, and at the same time, the upper lip rises,

f face

Structure

Exploring the
Construction of
Human Body Parts

Part 1

face

highlights and shadows ✏4

We use cosmetics to emphasize the three-dimensional aspects of our faces. For the parts we want to make more prominent, we use highlights or eye shadow. When we want to bring out depth, we use shadows.

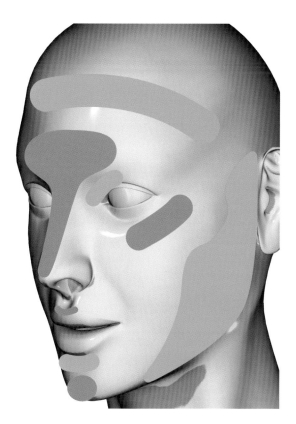

multiple layers of the "second skin" ✏5

Toner is applied to the face, followed by moisturizer and foundation. Also, face powder makes the skin's grain appear finer. The female face has many layers.

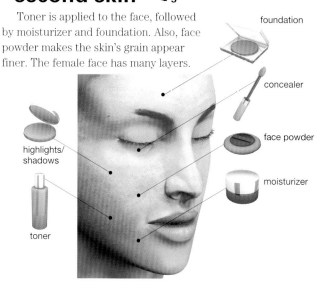

foundation

concealer

face powder

moisturizer

highlights/shadows

toner

cosmetics for selective emphasis ✏6

Cosmetics can be divided into two general groups: those that are used to hide trouble and those that are used to bring out features or texture, emphasizing or adorning certain parts.

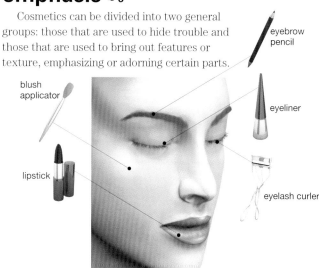

eyebrow pencil

blush applicator

eyeliner

lipstick

eyelash curler

revealing the front teeth. The nose also changes shape, and wrinkles appear from the wings of the nose to the mouth. When the mouth changes shape, the nose and cheeks follow suit. Skin stretches flexibly, but it doesn't shrink, so when one part is stretched out, there are bound to be wrinkles somewhere. When we laugh, the mass of skin on the cheekbone stands out, and wrinkles run down from both sides of the

mouth. However, skin will not always be pulled in the direction of the muscles. The mouth is never pulled directly to the side, and the eyebrows also don't move to the side.

✏4 ✏5 ✏6. The role of cosmetics

Women's skin is lighter and softer than men's. Cosmetics can be very important for women. They take exceedingly creative actions like

shaving their eyebrows, then using an eyebrow pencil to fill in the shape the way they like. Women use toner, first, to prepare the face; then moisturizer, which keeps the other cosmetics in place; then perhaps a concealer here and there to cover trouble spots; and then foundation, which is spread on the "canvas." To make the skin grain appear finer, they use face powder. They also create shadows with eye shadow and other

cosmetics, use highlights to draw out certain elements, and apply blush to the cheeks to give them more life. A woman's "second skin" is made of many layers. On top of all of this go eyeliner, lipstick, and mascara, which emphasize certain parts.

when the mouth is opened/closed ☞7

Open the mouth wide, and the jaw will stretch into the extra skin below. One of the special characteristics of the jaw is that when it opens wide, it creates a bulge just in front of the ear.

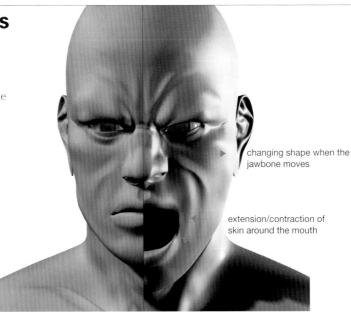

changing shape when the jawbone moves

extension/contraction of skin around the mouth

facial parts under motor control ☞8

Facial parts that look as if they move voluntarily include the sides of the mouth, the end of the upper lip, the wings of the nose, the base of the eyebrows, the upper and lower eyelids, and the frame of the lower jaw. Parts that move under motor control give the important impression of individual volition.

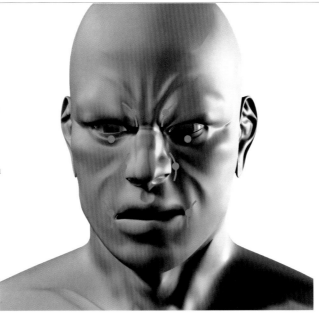

☞7 ☞8. Facial movements

The biggest movement in the face is done by the jaw, but unless we are singing or eating, the jaw doesn't move that much. When we talk, most changes are around the mouth. The eyes move in a more agile and lively way. Consider the wink, for example. Even with a small change in the eyes, the eyebrows move, and the cast of the eyes is set.

The softness of the cheeks is emphasized when the face moves. The cheeks stretch, gather, and form wrinkles—they work in concert with different expressions. They do not look like they can voluntarily move. Parts that look like they do move voluntarily include the upper eyelid, which unconsciously blinks. The lower eyelid sometimes moves in ways untied to emotion, but it is also important in creative expressions.

When the lower jaw moves in a big way, the shape of the head is also greatly altered. The movements of the lower jaw probably create the biggest lag between the bone structure and the skin.

All expressions are affected by emotions, and every activity occurs in a chain of actions. For example, we open our mouth wide when we yawn. We don't only open our jaw widely; the sides of our mouth rise in a diagonal fashion, and the upper lip is raised as much as possible. When this happens, muscles around the upper lip contract, and the eyebrows—connected at the other end—naturally are pulled down.

f face

Structure

Exploring the
Construction of
Human Body Parts

Part 1

face

facial asymmetry

We have to pay attention to the face's asymmetry when dealing with its shape and expressions. The two sides are not mirror images; even facial expressions have clear differences from the left side to the right. Depicting the face is key to infusing a character with life, which makes these minor differences of increasing importance.

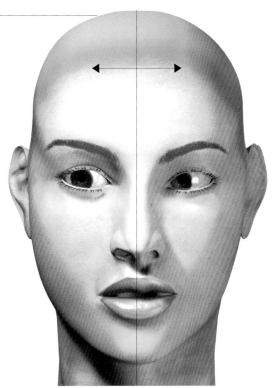

habitual expressions and postures ☞9

People have completely different facial expressions. The point is to bring out as much as possible the characteristics of each person's facial shape and look.

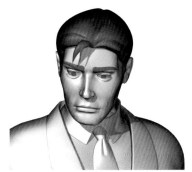

earnestly worrying

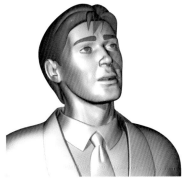

thinking lightly about something

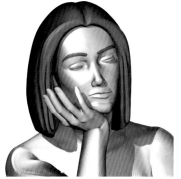

The skin changes because of the hand.

Peculiarities of the face and eyes share the same focus.

The face has many peculiarities. For example, when we are angry or we want to make our opinion clear, we look the other person right in the eye. The eyelids move in unison with the eye. The peculiarities of the face and eyes share the same focus in this instance. That's because the face, which takes in so much information, always goes toward the source of that information. If there is a loud sound, we turn our head toward it. If someone nearby is drawing our attention, we turn to that person. If danger is near, the face also tries to protect itself.

☞9. Capturing expression changes

Changes of expression are diverse, but when one does change, it changes all at once. Unless there is some special exception, the mouth doesn't go into a laugh before the eyes. Expressions transform in a hurry, like water pouring out of a cup. Of course, our expressions don't just affect our faces; they affect our whole body.

Expressions are not always beautiful, nor are they simple. Impassive looks, for example, create different impressions, and no two people have the same smile. The goal should be to bring out as much as possible the characteristics of each person's facial shape and look.

figure

figure

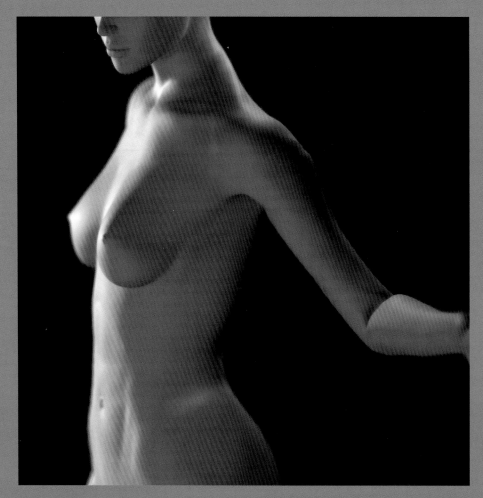

Figures change over time. In many Asian countries, for example, one big reason for this is the change to a more Western, high-protein diet. However, it is also caused by a general decline in muscle strength. For example, as muscles accumulate during a growth spurt, they compress the growth of bones. Looked at a different way, if muscles weren't involved with bone growth, bones would grow at a faster rate. This supposedly stylish and desirable appearance is not trouble-free. Muscle strength and physical stamina are tied to our spiritual condition. Without physical stamina, our drive and energy are difficult to maintain.

Related pages   Part 1 (all) Structure   Part 2 (all) action

f figure

Structure

Exploring the
Construction of
Human Body Parts

Part 1

figure

being well proportioned

When depicting the human body, choose from one of the methods for regulating its many proportions: the module, percent, or golden section method. The module method takes one section of the body and makes it a module, then uses that module to measure the whole body. A common module uses the head as the standard measure. The body is said to look best when drawn as eight head lengths, but it is actually seven head lengths plus a little extra. The percent method starts with the whole body equaling 100 percent. The golden section method uses a ratio of 1:1.618 to draw the human body.

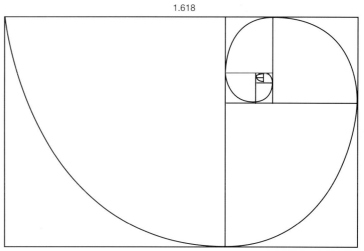

Connecting the apex of a rectangle with an inscribed line known as the golden mean

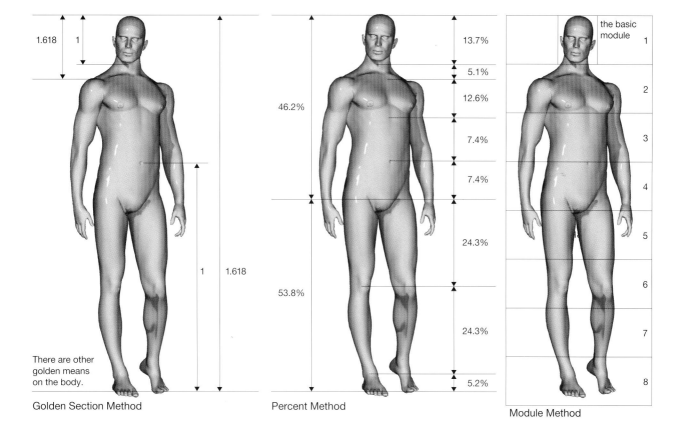

Golden Section Method

Percent Method

Module Method

sex differences in body shapes ☞1

When we enter adolescence, the male and female figures begin to differ sharply. Men develop hard lines from their muscles, while women develop rounded, pliant lines because they have more fat content.

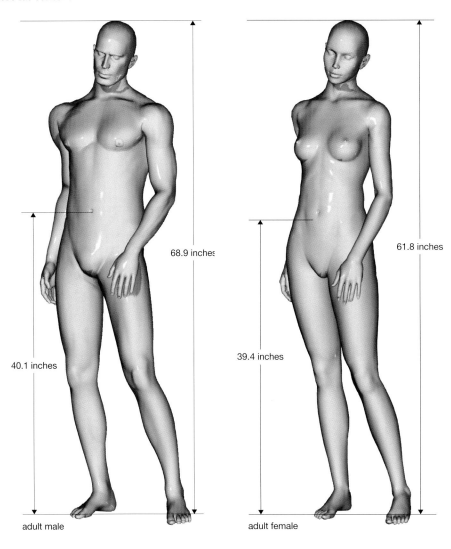

68.9 inches

40.1 inches

adult male

61.8 inches

39.4 inches

adult female

☞1. The complex muscular bulges of a man's body

Men are more muscular and have more bulges on the surface than women. Bulges appear at the base of the neck at the shoulder because of the trapezius in the back. In the arm, depending on which way the forearm is twisting, the outline of the muscle changes. Let the arm hang down and stretch out the fingertips, and they'll reach to the thigh. When you look down at the knee, you'll see that it is aligned slightly behind the center of the foot. Notice that from a side view,

if the man is standing, the heel is slightly behind the rest of the body. The parts thrusting forward the most are the chest or the belly. The shoulder blade juts out in the back.

☞1. The soft, flexible lines of a woman's body

Women are generally rounder and have softer lines than men. Their muscles don't really stand out. The hypodermis creates these soft lines. Women tend to be shorter than men; their hands and head are also

smaller. If men are about seven head lengths tall, many women are somewhere between six and seven. They usually have sloping shoulders, and the base of their neck doesn't bulge like men's do; it gently gives. With the exception of the feet, few muscle bulges can be found. Women's bone protuberances are similar to men's. Women's hips are bigger, and the shape of their buttocks and hips differs from men's.

age differences in body shapes

The body's form changes from infancy to old age. Below are drawings detailing the changes that happen at certain points in our life.

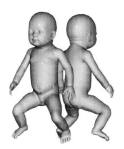

infancy
Infants have large heads and grow to four head lengths tall. The circumference of the chest is 11.81 inches; the circumference of the head is about 13.39 inches. Infants are anywhere from 19.69 to 31.50 inches tall. Their weight ranges from 6.6 pounds to 17.6 pounds.

adolescence
During this period, many parts develop quickly. The limbs grow flexibly. Females develop quicker and are practically full grown by the time they are 16.

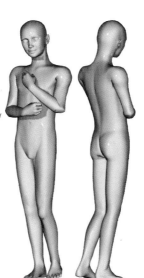

adulthood
This is the period when we reach our ideal body shape. Men develop broad shoulders, and their bone protuberances become more pronounced. Women develop balance as their chest fills out and their waist constricts.

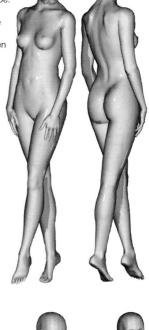

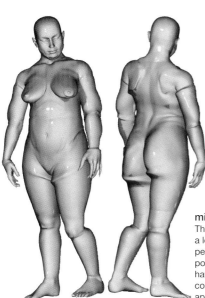

middle age
The body shape changes a lot in this period. Many people put on weight, possibly because they have become less concerned about their appearance.

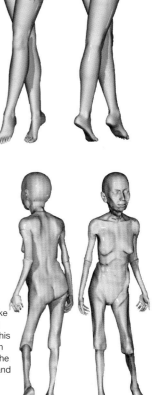

old age
Individual differences like thin and fat body types become more clear in this period than they were in middle age. Basically, the flesh thins out, drops, and becomes grainier. The face also develops deeper wrinkles.

☞2. The characteristics of an infant's body

Infants have little muscle strength, and their arms and legs are extremely short. But their body is little, too, so their legs and arms are actually proportionally bigger than an adult's. The backs of the legs are soft and plump, and an indentation can be seen around the skin at the joint. The abdomen is not constricted. The neck has yet to settle and is very short—it is hard to see at times.

Adolescence is the age when gender differences become striking.

As young women begin to grow breasts, their differences from men come to the surface. Because the muscle and bone structures have yet to be completed, the bulges on the surface are moderate. Young men's voices change. Also, their Adam's apple begins to show and their muscles get larger. In this period, the lines of male and female bodies become very different.

☞2. Putting on weight is easy in middle age.

Calories that were previously spent are now accumulated. Dietary habits have a big influence. Men tend to get fat faster than women. Creases develop around the eyes, mouth, and forehead. Fat develops at the bottom of the jaw, and flesh amasses around the lower ear. The shoulders and the back of the neck become more rounded. The muscles in the chest deteriorate, and wrinkles appear on the body. A woman's chest gets heavier. Flesh develops on both sides, and the upper arms get fatter, pushing the arms away from the sides. The belly becomes larger, and

the lower abdomen swells. The fingers swell; generally everything gets fatter. The waist stores excess flesh as it expands. The thighs also put on weight. The angle of the hip joint changes, spreading the feet farther apart. Some people can no longer cross their legs. The bumps on the legs begin to disappear under excess flesh. The ankle may seem relatively slim at this stage, but it is also getting fatter.

typical body shapes 3

Factors such as weight, height, fatness, and musculature change the way the body looks in motion or at rest. They change everything.

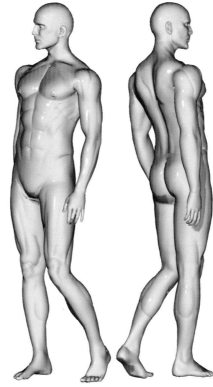

sinewy shape
The muscles are developed and the figure is well defined. Muscle bulges are noticeable, and they change remarkably during exercise.

obese shape
There is a surplus of hypodermis all over the body, and everything is rounded.

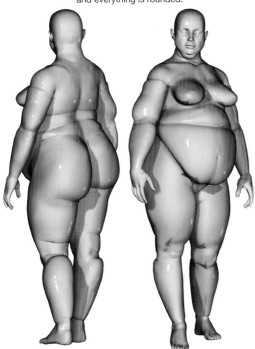

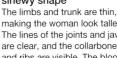

sinewy shape
The limbs and trunk are thin, making the woman look taller. The lines of the joints and jaw are clear, and the collarbone and ribs are visible. The blood vessels in the forearm visibly bulge against the skin's surface.

☛3. Moving with different body shapes

When two people have a weight difference of more than 45 pounds, their way of walking and many other movements differ. Differences even appear when they are both standing still. Thin people often wrap their legs and arms around each other—a habit that is developed; fat people stand with their legs and arms far apart— something they do out of necessity.

Height differences also create distinct characteristics. First, height decides a person's line of vision. Because a person hardly ever consciously contemplates how his line of vision differs from others, this difference is often overlooked. Yet a tall person often looks down at others and a short person often looks up.

☛3. Obesity produces changes in body shapes.

The flesh under the jaw obscures the jawline and hides the neck. The back of the neck also adds flesh. The arms fatten, making the joints seem smaller. The fingertips have a lot of excess flesh. The flesh from the chest extends to the area beneath the armpits, creating a thick layer. The belly sticks out and hangs down. The buttocks and back are also fleshy, but the fat is overwhelmingly in the front. The flesh on the back forms creases that head toward the front.

Fat builds up on the inside of the knees, keeping the feet from coming together. The body becomes wider and the limbs seem shorter, giving the impression that obese people are shorter than they actually are. Yet, the lasting impression is that obese people are big.

Excess flesh is definitely not light. In fact, it is very heavy. Heavy exercise becomes impossible, but the legs supporting all this weight become pretty fat themselves. All exercise is greatly transformed for obese people.

f figure

Structure

Exploring the
Construction of
Human Body Parts

Part 1

figure

units of length and body part measurements ☞4

Many things tie a body to its environment. Several units of length have their origins in body parts. Differences in height, which are standardized in many ways in our surroundings, create differences in action. The length of the thumb is measured in inches and the length of the foot is measured in feet. The meter is a comparatively new unit that dates back to 1875 when the distance between the North Pole and the equator was measured in meters.

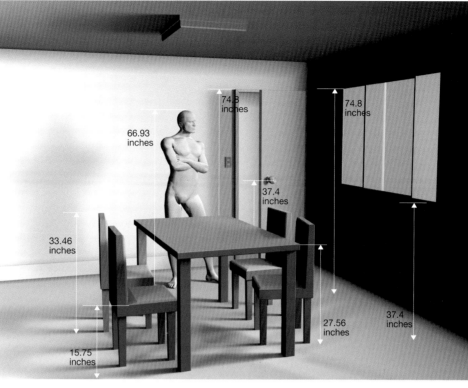

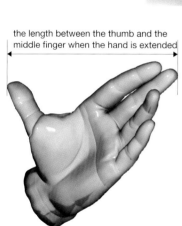

the length between the thumb and the middle finger when the hand is extended

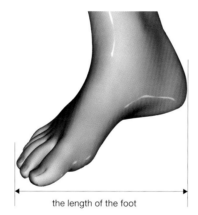

the length of the foot

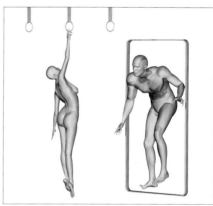

height and the environment

☞4. Body dimensions and their relation to the environment

Consider the way tall people pass through doors or how cramped they are when they get into a car. Short people often can't see the screen well when they take a seat in a movie theater. They may not be able to reach the hand strap on a train, and even if they can reach it, they may have their arm fully extended and get tired in the process.

When children are walking around for a long time, they can only pay attention to the range they are conscious of, which is why they tend to ignore the presence of tall people.

Le Modulor and human dimensions

The artificial world is made to fit the average person. Le Modulor, created by the famous French architect Le Corbusier (1887-1965), is a good example of this. Le Modulor is a methodology that determines the dimensions and length of the average human body through mathematical equations, and then uses that series of results in architectural structures. The method is just not for buildings—furniture and even open spaces are structured using this series of measurements. The idea was that this method would create spaces with a more fraternal feeling. But Le Corbusier was a large man, and he made the ideal height of a man 6 feet. The buildings based on this measurement must have seemed to the smaller Japanese people of that day like the houses of giants. For them, it would not have been a comfortable place to live.

Today, we reach for doorknobs and light switches without any discomfort because the planning is done in a uniform way to fit the human body. If there is an uncomfortable toilet seat in the house, it's a good bet it was not one of the products included in the planning.

shapes of action

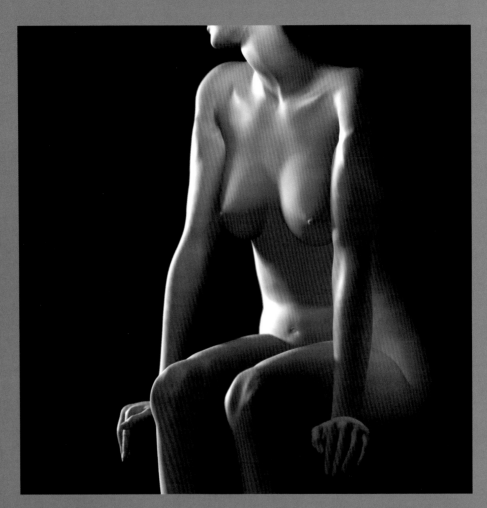

Move the body, and its contours change in many ways as the whole body, including the bulges of the muscles and the compressed flesh, transform. In what way does the substance that makes up the human body bend when the body moves? To understand the body, you need to reconstruct a simple model of its structure, learning anew where the hard and soft parts are. The shapes of the hard bones don't change, but the soft flesh that surrounds them does alter. As we move, our parts bend, which creates a ripple effect that sometimes even interferes with the bending of other parts. What shapes do our bodies form when we take action? Let's take a close look at some parts with special features related to action.

Related pages   p. 36 shoulders   p. 42 neck   p. 54 legs   p. 66 arms   p. 72 back & waist
Part 2 (all) Action

neck movements ☞1

Wrinkles form below the lower jaw when it moves downward. Open the mouth wide, and tension in the platysma as well as changes in other neck muscles can be seen. These are small movements that occur during intervals when the expressions are changing, but they shouldn't be overlooked.

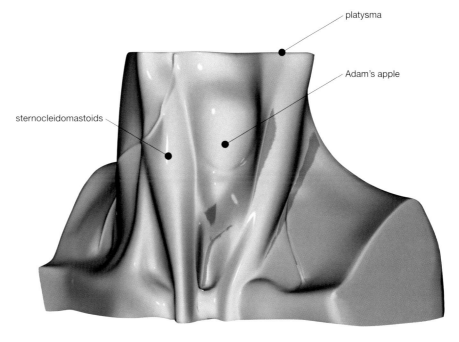

platysma

Adam's apple

sternocleidomastoids

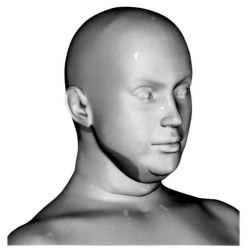

Superfluous flesh accumulates at the bottom of the neck.

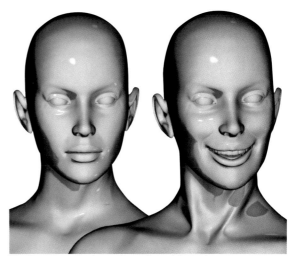

Mouth movements change the surface of the neck.

☞1. Large changes in the neck and Adam's apple

Large changes in the neck center around the sternocleidomastoids on either side, but the platysma below the jaw also affects expressions. It's especially apparent when someone laughs. The sternocleidomastoids change most when someone looks back. They are clearly defined by bulges that occur from behind the ear to the center of the collarbone.

The Adam's apple also changes the look of the neck. It moves up and down when a man drinks or vocalizes. Notice that the horizontal wrinkles on the neck can't keep up with the Adam's apple when it moves. The Adam's apple is pronounced on men, and because of the thin skin on the neck, even some of the Adam's Apple's small details are visible.

115

shoulder movements ☞2

Even when the shoulder only moves a little, the shoulder blade in the back moves in a striking way. That change is most apparent when we raise an arm.

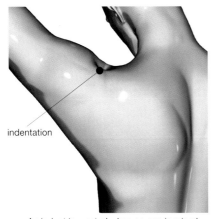

indentation

An indent is created when an arm is raised.

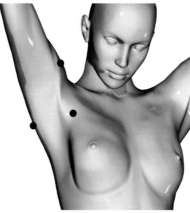

These spots are emphasized when the arms move.

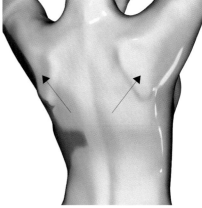

The shoulder blades move this way when the arms move.

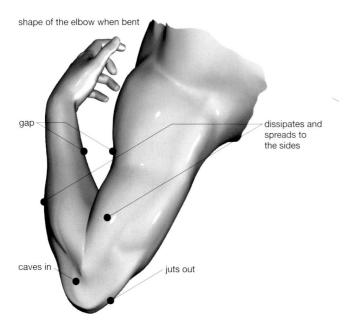

shape of the elbow when bent

gap

dissipates and spreads to the sides

caves in

juts out

This part of the upper arm expands when the arm is pressing on something.

☞2. Arm movements are accompanied by shoulder transformations.

The shoulder blade supports the arm as it moves diagonally upward and out. A large indentation is created between the collarbone and the humerus when the arm is raised. Extend the arm forward, and the pectoralis major in the chest tenses while the top of the shoulder bulges.

Notice that the collarbone and shoulder blade move along with these muscles. We can also see the flesh on the back of the armpit being pulled.

Upper arms have a lot of fat on them, which creates changes all over. Put your arms behind your back, and the flesh on the upper arms will bulge as it is pushed to the outside. Stretch the arms out a little to the

side, and the upper arms will slacken in an arc shape. Like the thigh, the parts that aren't tense are soft.

S
shapes of action

Structure

Exploring the
Construction of
Human Body Parts

Part 1

shapes
of action

flexing/extending the elbow

When the elbow is extended, the wrinkles draw near, and the indentations on either side of the bone can be seen. Flex the elbow, and the skin is stretched as the bone juts out.

wrinkles form here

indentation

skin between the thumb and index finger

The skin between the base of the thumb and the palm changes dramatically. It is linked to the movement of the thumb. Sometimes it looks wrinkled and tense, and at other times it can look like a web of skin.

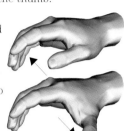

twisting the forearm

Two bones cross when the wrist rotates. Bulges appear in a corkscrew pattern on the arm, as if the bones and muscles were wrung out.

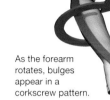

As the forearm rotates, bulges appear in a corkscrew pattern.

moving the hand

When we open and close our hands, the surface of the forearm goes through minute changes as the muscles below move. Make a fist, and the forearm from the elbow becomes thicker. Open the hand, and it becomes thinner.

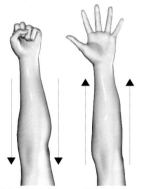

Relaxation slopes gently downward.

Tension moves up the forearm.

sex differences in hip shapes

When the legs stretch out, indentations appear at the center of both sides of the waist. The change is more striking in men. When a leg is lifted forward, a protuberance appears in women at the base of the leg on the upper front of the waist.

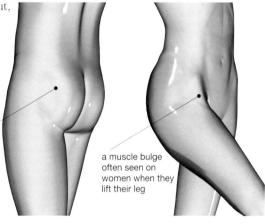

indentation seen on men

a muscle bulge often seen on women when they lift their leg

sitting bones and buttock shapes

When we bend the waist to sit, the thick flesh moves forward and the bottom of the pelvis juts out, making the buttocks look pointed.

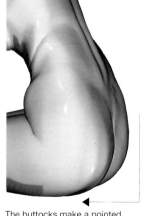

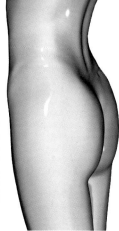

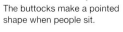

The buttocks make a pointed shape when people sit.

angle of the elbow at full flexion

The elbow can bend at an angle of 150 degrees at the most. At this angle, the wrist is about 2 inches away from the shoulder. In other words, there's no way for the wrist to touch the shoulder.

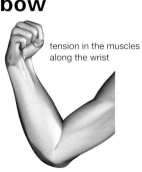

tension in the muscles along the wrist

skin folds in the abdomen

The belly usually has a lot of superfluous flesh. When the upper half of the body bends forward, more and more wrinkles approach the area around the navel. Excess flesh gathers in the lower abdomen, creating a strong line at the base of the leg.

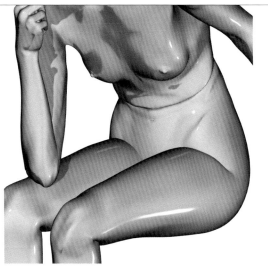

deformation of knees

When the leg is stretched, the kneecap forms a protuberance, but when it is bending, the kneecap moves lower and the bulging disappears, leaving the surface smooth.

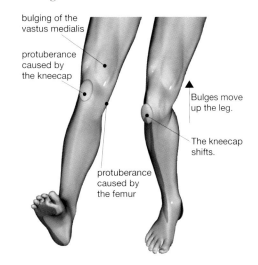

bulging of the vastus medialis

protuberance caused by the kneecap

Bulges move up the leg.

The kneecap shifts.

protuberance caused by the femur

adducting/abducting the legs

Open the legs wide and muscle bulges on the inner thigh are visible. Close the legs, and the flesh gathers in the middle of the thigh. If there is a lot of flesh, wrinkles develop on the crotch.

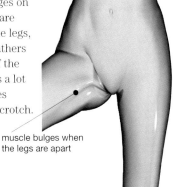
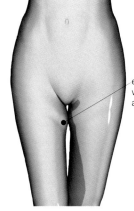

muscle bulges when the legs are apart

excess flesh when the legs are closed

deformation of legs in *seiza* posture

Sit in the *seiza* position, with your lower legs and feet tucked under your thighs and buttocks, and the thighs spread on either side. The top part forms an arc. The calves spread out in the same way. The indentations on either side of the knees disappear the more the knees are bent.

toes and tension

Extend your feet and your toes will naturally curl. Raise the toes, and you can see the tension in the muscles on the top of your foot. The surface of the shin also moves a little.

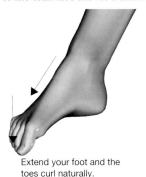

Extend your foot and the toes curl naturally.

backs of knees

Muscles bulge visibly on both sides of the back of the knee. Bend the knee, and the muscles become tense. Many wrinkles appear here.

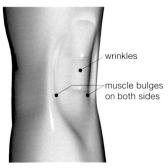

wrinkles

muscle bulges on both sides

back of the knee

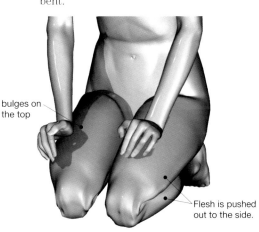

bulges on the top

Flesh is pushed out to the side.

coordinated movements of body parts ☞3

Raise the arm straight up, and the chest will tighten. To rotate the arm even higher, the spine helps out by bending backwards.

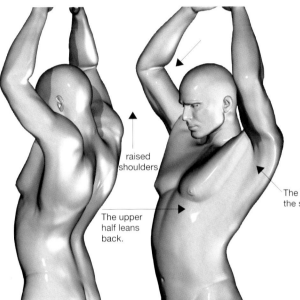

raised shoulders

The upper half leans back.

The head nears the shoulder.

the way the body transforms when raising an arm

back muscles stretch

hip line rises

calf rises

rise of the heel when wearing high heels, etc.

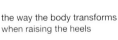
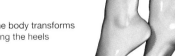

the way the body transforms when raising the heels

☞3. Bulges on the ribs when an arm is raised

Begin to raise an arm, and the spine will bend backward. At this point, parts of the rib cage become visible as they are being pushed out. The muscles in the back tense up, protruding in an area above the waist. Even if the head doesn't move, when the arm is raised, we instinctively draw in our jaw a little. The collarbone moves inward; the

shoulder blade moves to the outside; and the shoulder moves all the way to the side of the head.

☞3. Movements are all related.

The high heels example above is one of many that show how movements are linked. Even little movements are connected in complex ways, and ignoring those connections will result in movements that look unnatural. Gravity is an

important characteristic of movement that should not be forgotten. Also, remember that when moving a part to the limits of its range, other parts will help it out.

When we take one action, other actions are often taken to help the first one. When we get up, we stand on our legs and use the weight of our legs to lift our torso. We react as we rise. When we jump, we extend our arms backward and swing them out,

making a higher jump possible.

Actions largely occur in three phases: preparatory actions, the main action, and follow-up actions. There are also times when preliminary actions—well before the three action phases—and adjustment actions, which help bring the body back to a normal posture, are called for.

119

Part 2

Action

Exploring the Mechanisms of Body Movements

Once you've grasped the structure of the body, it's time to move to the excellent teachings of animation production on the subject of human movements. Capturing these movements is not the work of cell animation, with its simplistic sets of signals. It's very complex, and each little movement is linked to create bigger movements. Also, it's not about capturing the most superficial aspects of movement. It's about finding the reason for the movement–is it a physiological reaction, is it a spontaneous reaction, or is there no motive at all? In Part 2, we will introduce and practice many basic movements.

Note

In this part, we will refer to time-ordered movements as "frames." When we introduce simple variations, we will call them "patterns." When we record human movements on video, 1 second equals 30 frames. Frame 0 is the time just before the movement starts. Also, please note that "pattern" has no time-related meaning.

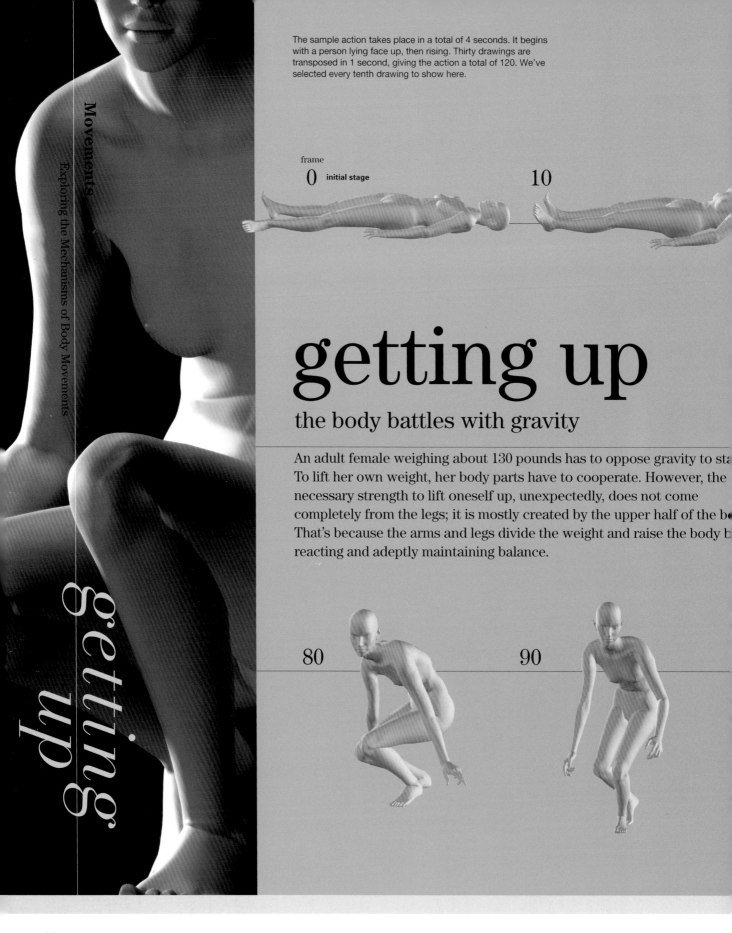

The sample action takes place in a total of 4 seconds. It begins with a person lying face up, then rising. Thirty drawings are transposed in 1 second, giving the action a total of 120. We've selected every tenth drawing to show here.

frame

0 **initial stage**

10

getting up
the body battles with gravity

An adult female weighing about 130 pounds has to oppose gravity to sta
To lift her own weight, her body parts have to cooperate. However, the
necessary strength to lift oneself up, unexpectedly, does not come
completely from the legs; it is mostly created by the upper half of the b
That's because the arms and legs divide the weight and raise the body b
reacting and adeptly maintaining balance.

80

90

getting up

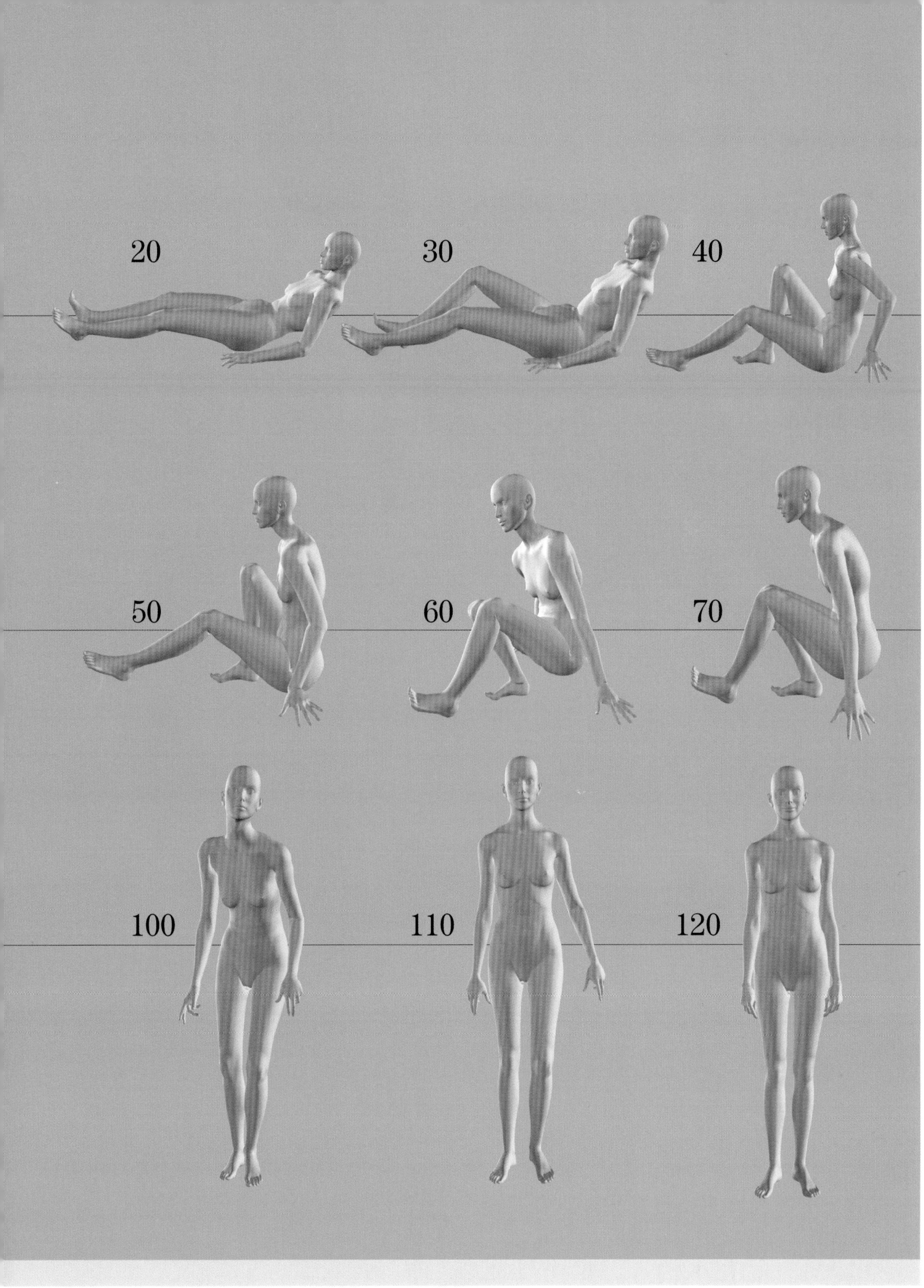

g

getting up

Movements

Exploring the
Mechanisms of
Body Movements

Part 2

getting
up

20

30

40

50

60

70

100

110

120

0

lying down ☞1

The action starts with a person lying face up and still. The body, which is not flat, shifts around to create as many points of contact with the ground as possible.

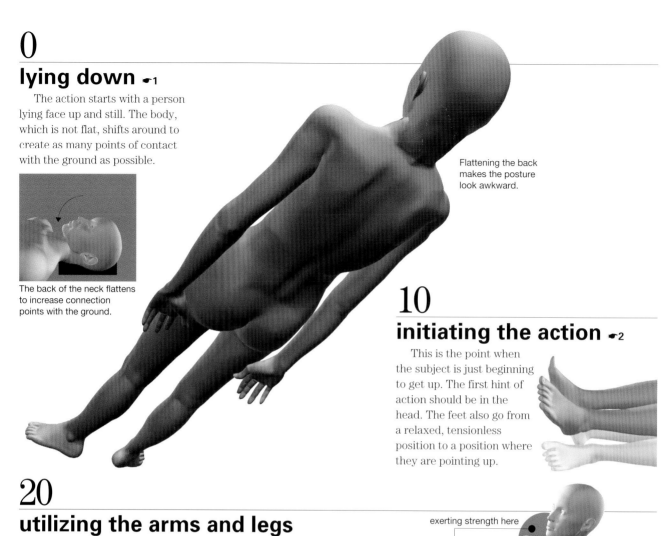

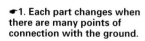

The back of the neck flattens to increase connection points with the ground.

Flattening the back makes the posture look awkward.

10

initiating the action ☞2

This is the point when the subject is just beginning to get up. The first hint of action should be in the head. The feet also go from a relaxed, tensionless position to a position where they are pointing up.

20

utilizing the arms and legs

The stomach creates layers as the body rises. The elbows begin to bend, supporting the weight of the upper half of the body. The buttocks become the point of support, while the feet hover above the ground. The jaw is also more drawn in now.

exerting strength here

support

☞1. Each part changes when there are many points of connection with the ground.

The back of the head is not flat. To make it flatter, the back of the neck to the middle of the head is stretched out. The chin is drawn in. To make the shoulder blades lie comfortably, the shoulders are slightly shrugged. To get the elbows pointing down, the elbow side of the forearm makes contact with the ground, which might push the shoulders toward the chest.

When women lie flat on their backs, their breasts fall to either side. The palms of the hands lie facing the ground, and the forearms are turned inward. The stomach sticks out farthest on the torso. The neck, shoulders, buttocks, and legs make contact with the ground, allowing the spine just above the hipbone to naturally curve. The legs from the heels to the calves are flat and facing the ground. Both legs tilt toward the outside and are slightly spread apart.

The knees are slightly bent. The toes are extended comfortably.

☞2. Initiating the action

You can't raise the upper half of the body without first lifting the heavy head. At this point the legs and other parts do not pitch in; only the muscles in the neck and waist help lift the head. As the back rounds, the top of the waist dips a little. The shoulders shrug slightly, which may express the body's desire to keep rising. Usually,

the eyes are focused just beyond the legs. This comes from the habit of staring at where the head is going. The body is no longer relaxing. You can see the tension in the toes and fingers. The toes extend upwards, while the fingers are slightly rounded as they prepare for the next action.

☞3. ☞4. Using body weight

The body has moved to a point where the elbow can support the upper half. The arms are gathering

g
getting up

Movements

Exploring the
Mechanisms of
Body Movements

Part 2

getting
up

30
shifting the body weight ☞3

The legs, which looked like they were about to float into outer space, are bent. One is especially bent, stabilizing the lower half of the body. By making the body shorter, the legs shift the center of gravity higher.

40
stabilizing the balance ☞4

As the upper half of the body rises, the center of gravity drops. The arms brace for the next action as the hands touch the ground.

50
preparing for the thrust up ☞5

The arms and the rest of the body coil like a spring to create the vigor needed to lift the body up.

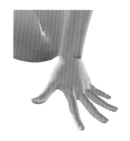

60
insecure departure ☞6

At first the legs are floating. Then most of the body weight is transferred to the right leg. Both arms are also helping out. To maintain balance, the left leg quickly bends back to be closer to the waist.

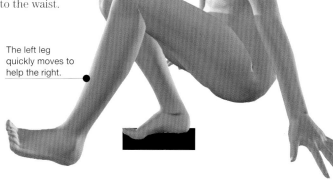

The right leg decides on a spot to settle and support the body weight.

The left leg quickly moves to help the right.

---

strength, but the legs have yet to go beyond bending. The bottoms of the feet are about to push against the ground. The hands are doing little at this point; the burden falls to the elbows. As the body weight shifts, the neck is still exerting strength.

☞5. Preparing for the thrust up

The body is almost vertical; the back is rounded. The arms exert themselves to shift the weight of the body onto the right foot, and the

hands press on the ground. However, the palm doesn't do the pressing; the strength of the hand is amassed in the five fingertips. The forearm tenses up as the fingers exert themselves. The head tilts slightly down. The shoulders spontaneously rise a lot. The left foot also shows a little tension. The right foot adheres to the body, and the heel is firmly planted on the ground below the waist. This is the instant when the person begins to push herself off the

ground.

☞6. Insecure departure

The right hand begins to move forward to counteract the coming shift in balance. Both legs are straining now, and the bulging muscles are clearly apparent. The right leg is fully folded, which means the thigh is not that strained, but the calf and Achilles' tendon are working hard. The upper half of the body is about to propel itself forward. The

right foot has already settled on a position for supporting the body, and the right hand is also supporting the upper body, but overall, the body is still unbalanced. Compared with 10 frames earlier, the left leg has moved quite a bit. The hips have been lifted just a little, but we are getting a sense of just how difficult it is to do this. This is the most dramatic stage of the action.

70

ready for the action ☞7

We've seen the first half of the action up to now. Both feet find their places and support the weight of the body.

90

balance secured but posture still not

The body weight is nearly stabilized, and the feet are fully supporting the whole body. The right foot is slightly behind the left with the heel raised.

80

body starting to get up ☞8

The waist separates from the feet as they begin to lift the upper half of the body. Both arms try to maintain the body's balance.

100

now standing but still awkward

The posture is still incomplete. The left foot is supporting the body weight. The arms grope for a comfortable position. The right foot is slightly off the ground.

The body leans as it returns to a standing position.

☞7. Ready for the action

The right hand is extended forward to maintain balance. The left hand is still pressed to the ground to keep the body from falling to the side. The arms are fully extended, but they are not exerting that much strength. The shoulders are slightly shrugged and leaning forward. The jaw is jutting forward as much as possible. The line of vision has dropped to the ground. The legs and stomach have a little leeway; they are resting before they begin the next action.

☞8. Body starting to get up

As the body starts to get up, the left hand leaves the ground. The left leg is very strained, and the right leg is supporting it. The thighs are also straining. The back is straight; the shoulders are raised; and the protuberances on the collarbone are clearly visible.

☞9. Final adjustments

The waist and upper body have moved back. The left arm looks as if it has been thrown to the side. The right

arm moves in the same way, but it stops close to the body. The back muscles straighten as the upper body rises.

☞10. Fixing the posture

The arms, having been flung aside, consciously come back to the body. As the back muscles stretch, the shoulders relax and gently slope downward. The neck is extended and finally holding the head in a fixed way. The center of gravity has returned to the middle of the body,

and the knees are slightly bent the other way as the legs stand straight. The standing posture is complete. It is different from the previous positions in that it can be held for a long time. It's comfortable. The action is completed.

Becoming dizzy when you stand

When you stand up fast, you sometimes feel dizzy or get a headache. In this case, one hand begins to search for something to

g
getting up

Movements

Exploring the
Mechanisms of
Body Movements

Part 2

getting
up

110
final adjustments ⟢9

It's time for the final
adjustments. The right leg is still
held aloft. The waist has settled into
its final position.

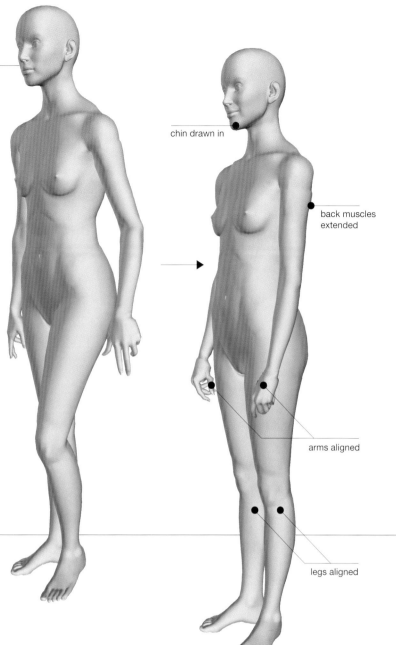

chin drawn in

back muscles
extended

arms aligned

legs aligned

120
fixing the posture ⟢10

After standing, the final step is to
fix the posture to enhance one's
appearance. The right leg is now
aligned with the left.

grab onto, while the other presses
against the head. We see this more in
women, who tend to have lower
blood pressures than men, and very
old people. When humans went from
being quadrupeds to bipeds, the head
was more than twice as high. This
caused some big reactions. For one,
dogs and other quadrupeds don't
suffer from vertigo or dizziness like
we do. Most people can remember
getting into a car or roller coaster
and feeling sick, as your innards
seemed to float. The restrictions of

gravity affect our inner organs and
our blood flow, showing us anew how
hard it is to oppose this force.

**Keep in mind the following
movements**

Getting up is a visual activity that
hints at more to come. As a sample,
we've covered an action that begins
and ends in still postures, but
generally the person will begin to
walk or commence another activity.

The weight of the body's different
parts can feel like shackles at times,

but when used correctly, these parts
can also make exercises easier.
When our posture is low, our arms
reflexively push against the ground
and spread to the sides. If we aren't
standing in a perfectly erect way, the
feet will move forward or back to find
the best way to maintain balance.
This allows the legs to maintain
balance in all directions. The feet,
which are the basic supporters of the
body's weight, adhere to the ground,
or one of them has its heel raised in
anticipation of the next move.

Keeping balance while trying to rise
is difficult, which is why this is a
disorderly movement. But the basics
don't change: The head looks up; the
stomach muscles raise the upper
body to a certain degree; the arms
kick in; the weight of the hips is borne
by the legs; and then the legs raise
the whole body. When stringing the
act of getting up with another action,
remember to consider the way the
actions influence each other.

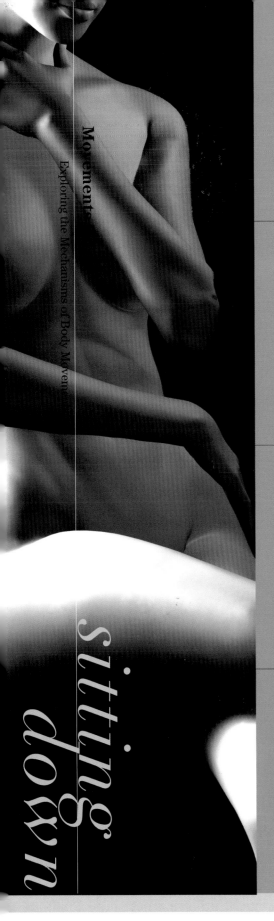

This sample action is 2.8 seconds long. The person stands up and sits down again in the same "chair." Thirty drawings are transposed in 1 second, which means there are 84 frames in this action. We've pulled out every seventh drawing here.

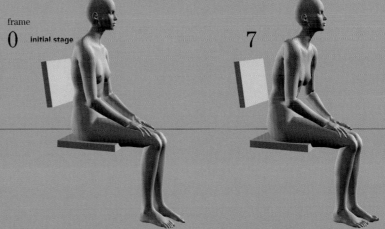

frame

0   **initial stage**

7

sitting down

placing the body

Sitting in a chair is an action that depends greatly on the type and size of the chair, as well as the environment of the person doing the sitting. Also, the act of getting out of a chair does not mirror the act of sitting. Think about the body, bound as it is by gravity, and it's clear the two actions would not be similar. In this chapter, we'll explain the acts of getting up from a chair and sitting down again.

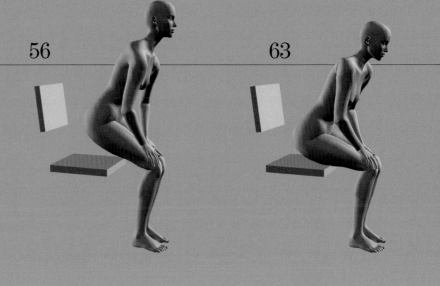

56

63

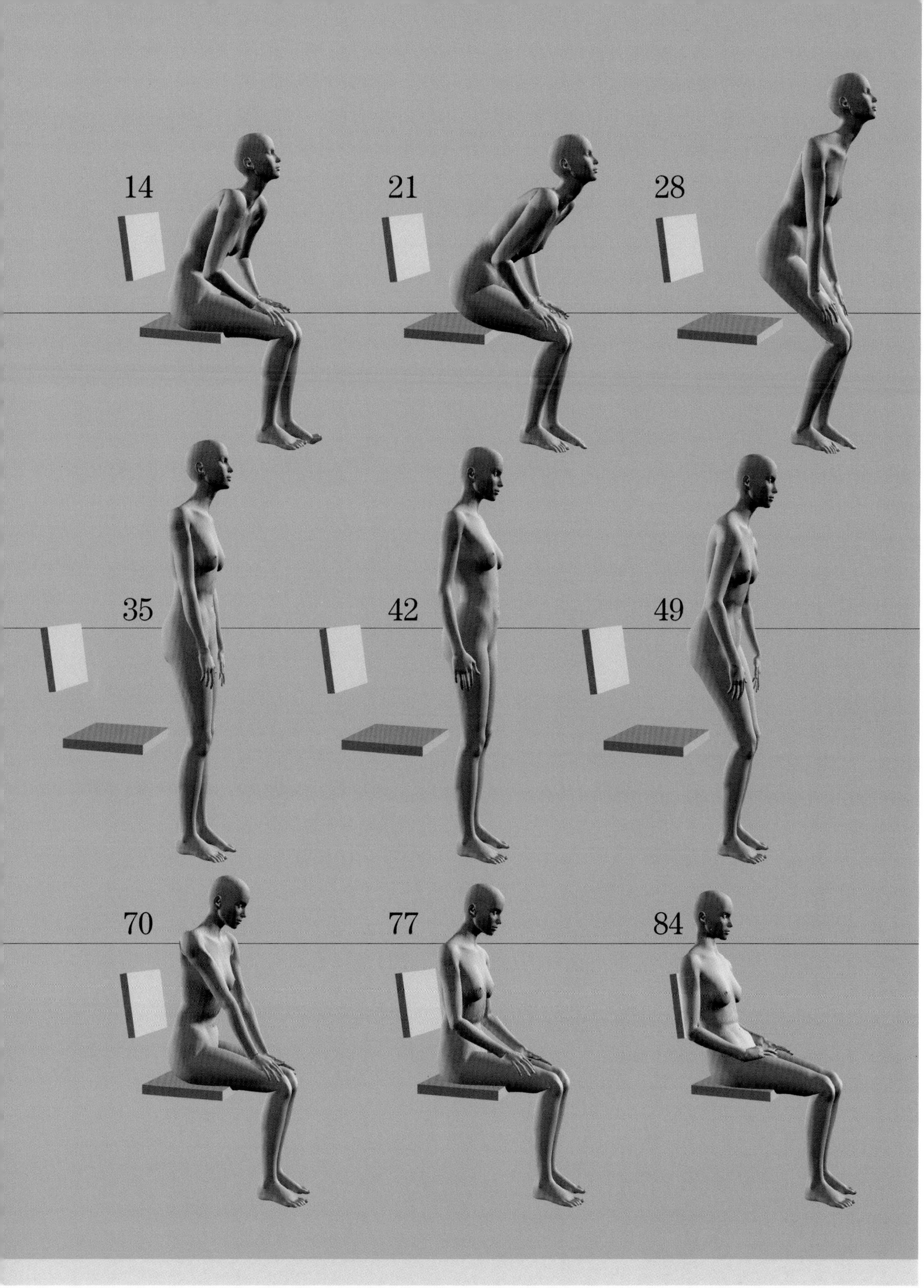

S
sitting down

Movements

Exploring the
Mechanisms of
Body Movements

Part 2

sitting
down

14

21

28

35

42

49

70

77

84

0

sitting postures ☞1

All of the body's weight is deposited in the chair. The buttocks are flattened out. Excess flesh hangs down from the thigh, which is just off the chair seat. The hands have nothing to do, so they rest on the thighs to stay out of the way. The shoulders droop. When the back muscles are stretched out, the head juts forward a bit, and when the upper half of the body leans against the back of the chair, the neck should be slightly raised. The legs have no functional role to play, so they take a variety of positions. For example, the legs can be crossed or leaning to the sides, and sometimes one is tucked under the buttocks.

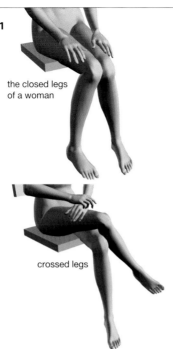

the closed legs of a woman

crossed legs

7

initiating the action ☞2

This is when the person is about to stand. The bottom half of the body has yet to change, and the upper half is just leaning forward a bit. This act of bending forward doesn't require strength; it can be easily restrained by a single finger against the forehead. In other words, if the head doesn't move first, the upper body can't lean forward.

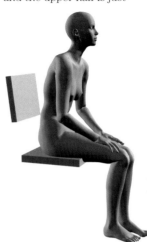

Before a person stands, the upper body leans forward.

14

shifting the body weight

The back is slightly stooped. Note that the spine, the gathering point for a succession of small bones, creates a gently sloping curve.
The upper body is leaning fairly far at this point, creating layers of flesh on the stomach. The jaw juts forward quite a bit. The hands can't go any farther forward, so the elbows bend and the hands keep their place just before the knees.
The person begins to exert herself as she prepares to rise. The arms stretch a little to the sides. The legs are still not straining.

21

departing from the chair

At last, the hips part company with the chair. This posture—filled with vigor as the body reacts—occurs for just a fleeting moment. The important point here is that the head and the parts in contact with the ground—the heels—are vertically aligned. If the head goes past this line, the center of gravity will be thrown forward and the person will tumble. From a side view, the body forms a "4." The arms are straining as this action proceeds.

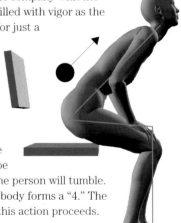

☞1. Sitting postures
Our posture often slopes. Even when we straighten our backs, the shoulders are rounded and pointing inward, and the weight of the arms is placed on the ribs. The chest faces down. When the legs are crossed, they practically form an arch. The calf of the leg on top is stretched and swollen, and the toes are extended naturally. If someone sits in a chair without much exertion, the legs will open slightly to either side, and the toes will inevitably point toward the outside. Women wearing skirts or wanting to look refined put their knees together and close their legs. In this case, the heels are slightly spread apart.

☞2. Energetically getting up
To maintain a level field of vision, the chin juts out a little. When people are sitting deep in a chair or when they are tired but try to stand anyway, they tend to have a bigger reaction, leaning forward more. The face follows and looks downward. The shoulders move forward, and the hands, resting on the thighs, slip ahead. But the shoulders and other parts aren't exerting themselves. If a person can't lean at the beginning, standing becomes difficult.

S
sitting down

Movements

Exploring the
Mechanisms of
Body Movements

Part 2

**sitting
down**

28
straightening the limbs 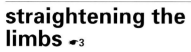3

The legs have almost successfully completed the task of supporting the body weight. All that's left is for the upper body to straighten up. The body formed a "4" before; now it forms an "S."

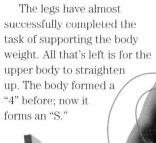

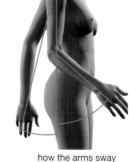
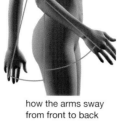

how the arms sway
from front to back

bending in an "S"
shape while
maintaining equilibrium

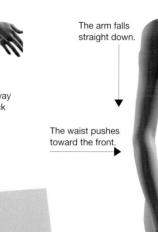

The arm falls
straight down.

The shoulder is
pulled back.

The waist pushes
toward the front.

the standing
posture

35
putting things in order

The body weight has been stabilized, and the feat of standing has basically been accomplished. Some of the joints are still a little bent, and the chin is still thrust out. The back muscles are not yet fully stretched out. When someone stands up with force, the back bends backward, the knees are more bent, and the chin is pulled in.

42
finishing the action

The chin is drawn in; the back is extended. The knees are also fully extended. The body has reached its final position, but from the side, it looks like it's slightly curving. The arms are arranged at the sides. Standing is exercise for the body's frame. It's a posture with a lot of special characteristics, and we need to probe enough to know which muscle is tensing up and what is being supported by the bones.

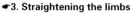

3. Straightening the limbs

The arms have finished their role; they hang at the sides. When rising energetically, the hands go toward the inside. Once the body has risen, the hands bump into each other in the middle—briefly forming a half circle, and then returning to their respective sides. Even when they don't hit each other, the hands cross over each

other and return like a pendulum.

Often, body parts that aren't being used in a certain action tend to be disguised as just so much heavy matter. But reactions, habits, and related physical images are very important characteristics when it comes to expressing movement. Information about the feel, weight, thickness, and range of body parts is

essential.

When creating images of people swaying their forearms while standing, keep in mind that when the arms swing forward, the emphasis is almost all on the action of the forearm. The elbow is slightly bent at this point, but when the arms swing back, the elbow straightens. At this point, the forearm's range of

movement is complete, and the shoulder continues to swing the whole arm back.

49

going insecurely for the seat ☛5

Time to sit down again. Before doing this, a person would definitely check the chair. At this point, most of the body weight is still supported by the feet, so the person can still stop the action.

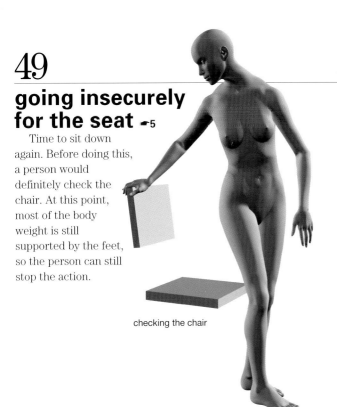

checking the chair

56

letting the body fall ☛6

The shape of the body is similar to the "4" when standing up, but the parts that are straining and the direction of the movement is completely different.

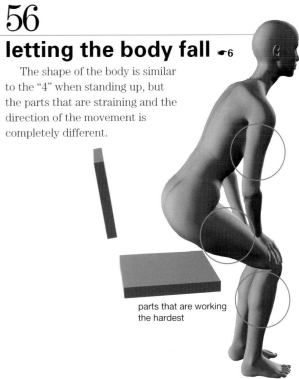

parts that are working the hardest

63

finding the seat ☛7

At this point, the hips haven't decided exactly where they will sit. Usually they go for a preliminary spot, rather than the one they will later settle into. The person's weight has not yet hit the chair, but, even with exertion, the body can't rise again.

The head maintains balance.

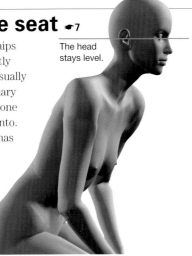

The head stays level.

70

getting seated

The person has now been seated. The whole body quickly relaxes. The body is still leaning awkwardly. The hands are still on the knees; they have nothing to do. Once the legs relax, the weight rests on the waist, and the buttocks and thighs are flattened out.

77

finishing up

The full weight of the body is in the chair, and we are back to the starting position. The back is straight and stable. The legs are relaxed. Both arms are in front. The waist and hips are supporting the upper body.

☛5. Before sitting in the chair

Before sitting, the person looks at the chair. She checks because she'll fall on her backside if there is no chair. When we stand, we sometimes do so in an energetic way. This is never the case with sitting (with the exception of sturdy sofas or other chairs we can jump into without falling).

☛6. Sitting posture

When sitting down, the chin, which has been wholeheartedly thrust out, is pulled in a shade. The arms also strain a little, and the muscles in the legs are tense. When the weight of the body is on the heels, the toes are suspended in air or pointing up. The muscles in the thighs are tense as they work to steady the legs. The back is bent.

☛7. Head movements when finding the seat

The head cautiously stays level; it's built to adjust quickly if the body should lose equilibrium. The head is not only an information processor, but also a balancing device at the top of the body. If you turn your head when standing, your body will naturally follow. When the head moves, the waist works as an axis to counter that move.

Actions differ from chair to chair

The actions of getting up from a chair and sitting down resemble each other, but they are totally different. The body lines are also different. When standing up, we try to stand in an energetic way. When we sit down, we try to eliminate any overly energetic movements. Getting up and sitting down are very frequent actions, and of course, they are very important. Note that the actions are

S
sitting down

Movements

Exploring the
Mechanisms of
Body Movements

Part 2

sitting
down

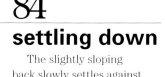

sinking down

84
settling down

The slightly sloping back slowly settles against the chair. The head is slightly thrown back and the chin is slightly drawn in. The hands are put in a place where they won't be in the way, such as next to the waist. The legs part a little to either side. The person can hold this posture for a long period, possibly crossing her legs from time to time.

ETCETERA

Sitting in a seat that doesn't fit the body

The way we sit depends on the type of chair. In a low chair, the knees are higher than the waist, and it's hard to shift the body weight accordingly. In a higher chair, it is impossible to fully relax the stomach and/or leg muscles.

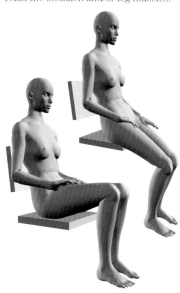

strongly tied to the type of chair involved. With a low chair, the body weight has to sink quite a bit; managing that weight shift is difficult. With a tall chair, the waist and the body weight must be lifted. If there is something on the chair to grab onto, it is easier to sit this way, so most people will use it.

Note that actions match the chair involved, whether it's a sturdy sofa, a tubular chair, or a chair on rollers. Of course, it is not all up to the chair; we must also pay attention to the person. If she is wearing a long coat or a skirt, she will tuck it in behind her before sitting. The person's mood will also alter the action, depending on whether she is tired, thinking, or in a different state of mind.

Linking to the next action

The way we get up depends on our next action. For example, if someone is in a hurry, she will first lean way back, then burst forth from the chair, propelling herself into the next action. When someone is reluctant to get up, that person may lean sharply forward, raise the waist only a little, and continue on with the shoulders slumped.

Our posture changes according to how long we sit

If we have a completely relaxed posture when we are asleep, then we have a half-relaxed posture when we are sitting. There is a strong likelihood we will stay seated for a long time, or will perform tasks while seated. Sitting still for a long time is difficult, and maintaining the same posture is painful. If the person sitting is not asleep, we're bound to see some sort of change in the posture. When depicting any activity done while sitting, we need to be clear on just what parts of the body are still and what parts are moving freely.

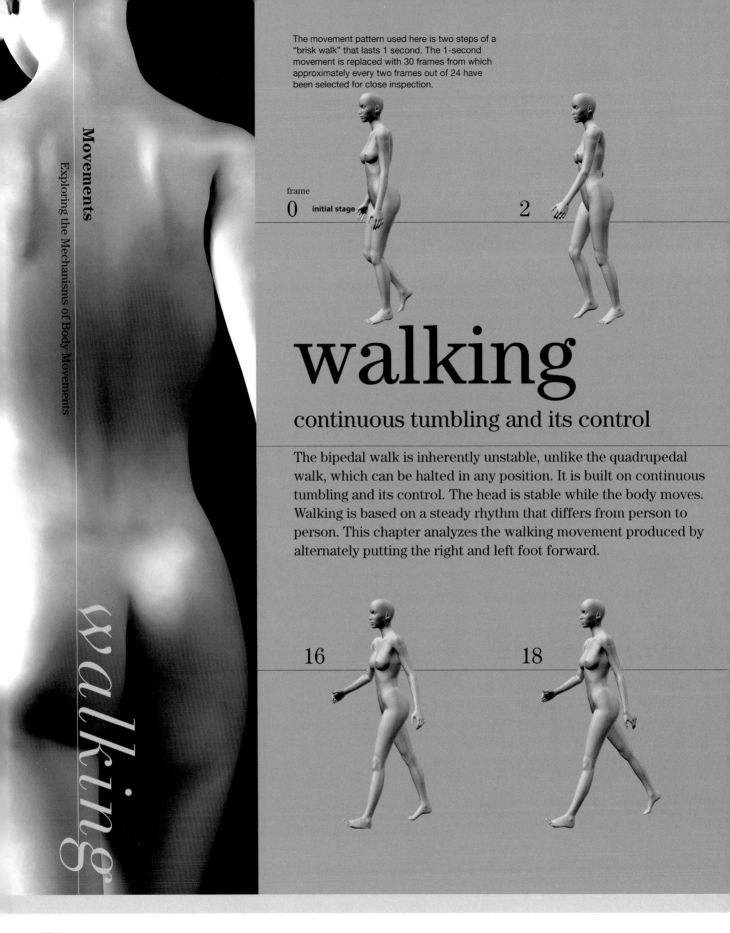

The movement pattern used here is two steps of a "brisk walk" that lasts 1 second. The 1-second movement is replaced with 30 frames from which approximately every two frames out of 24 have been selected for close inspection.

frame

0   initial stage

2

walking

continuous tumbling and its control

The bipedal walk is inherently unstable, unlike the quadrupedal walk, which can be halted in any position. It is built on continuous tumbling and its control. The head is stable while the body moves. Walking is based on a steady rhythm that differs from person to person. This chapter analyzes the walking movement produced by alternately putting the right and left foot forward.

16

18

walking

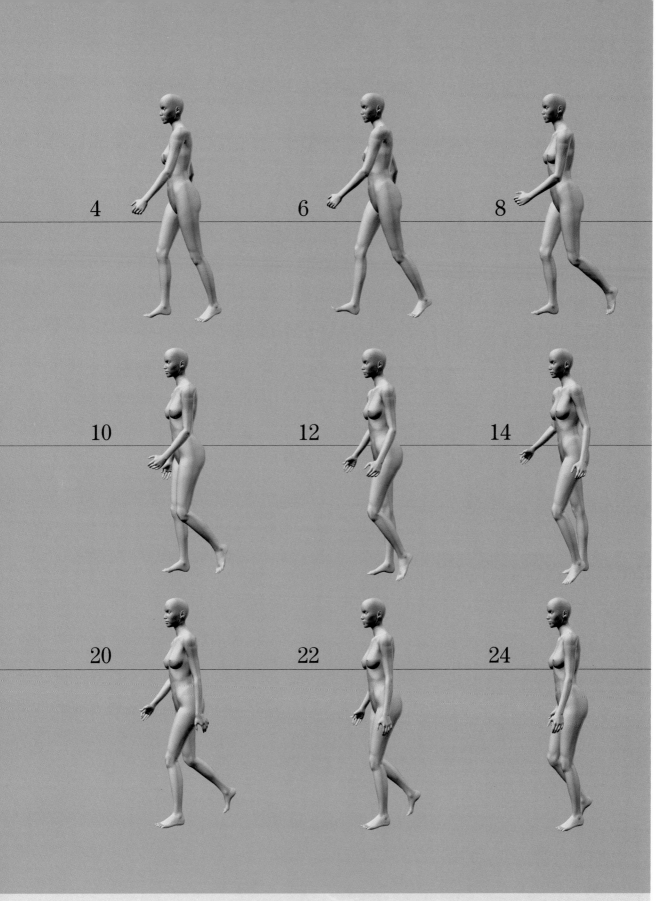

W
walking

Movements

Exploring the
Mechanisms of
Body Movements

Part 2

walking

4

6

8

10

12

14

20

22

24

0
sending the right leg forward

Walking is a steady repetition of tumbling and control. The position starts the moment the right leg is sent forward. The head faces directly forward and the arms hang loosely. The left leg tenses and supports part of the body weight. The right leg is bent and brought forward. The left leg's hip, which forms the fulcrum, is positioned higher and faces to the right. A crease forms below the left buttocks as shown by the right side's extension. At the same time, the spine and head twist laterally to keep the head level. The upper body faces and bends forward, just before the center of gravity is thrown forward and the balance is upset.

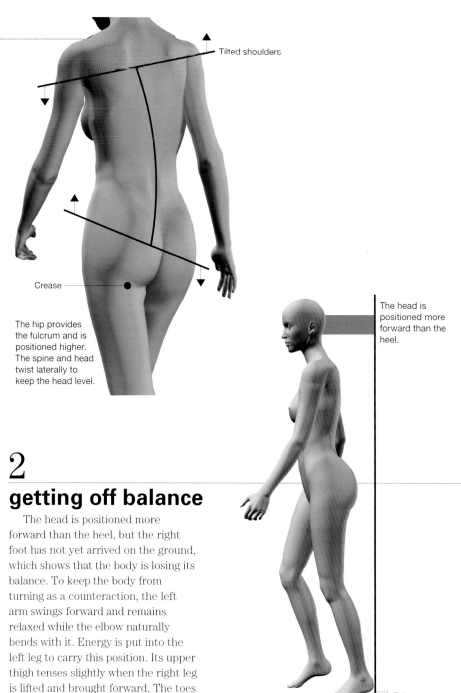

Tilted shoulders

Crease

The hip provides the fulcrum and is positioned higher. The spine and head twist laterally to keep the head level.

The head is positioned more forward than the heel.

Fulcrum

2
getting off balance

The head is positioned more forward than the heel, but the right foot has not yet arrived on the ground, which shows that the body is losing its balance. To keep the body from turning as a counteraction, the left arm swings forward and remains relaxed while the elbow naturally bends with it. Energy is put into the left leg to carry this position. Its upper thigh tenses slightly when the right leg is lifted and brought forward. The toes are flush against the ground.

The walking motion

Bipedal walking allows human beings to stand upright and walk on two legs, while the more stable quadrupeds have four fulcrums. The human spine creates efficient movement in walking or running—we are able to stand easily and run faster than animals—but we do get backaches. So why did we become *Homo erectus?* An erect posture supports a heavy brain that enlarges when greatly cushioned by the spine.

The walking rhythm

Walking is based on a steady rhythm: alternately putting the right and left foot forward to create a two-beat cadence. Adding a "stop" produces the three-beat waltz cadence. Such rhythm differs from person to person.

(W) walking

Movements

Exploring the
Mechanisms of
Body Movements

Part 2

walking

4

advancing the right leg

The right leg with its mobile lower part below the knee is propelled forward, but has not yet landed on the ground. The greater the degree of fatigue (careless motion) or hurriedness (the leg quickly pushes forward), the more prominent the kicking motion becomes. While the left leg remains still, the heel is lifted up as the body and arms advance forward. The right arm swings in reverse to keep the upper body from twisting.

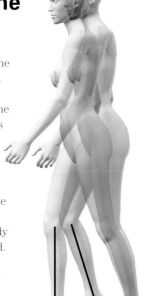

Advancing the leg with a kicking motion

8

landing and its impact

The body weight is almost relieved from the left leg, and borne by the right leg, which is firmly planted on the ground. The sole of the sliding foot lies flat and momentarily bears the weight to allow the left leg to bend at the knee, which is perpendicular to the ground. The right foot's large impact on the ground generates a brief vertical jolt in walking. The chest, buttocks and other areas with excess flesh jiggle as an aftereffect. A woman's chest momentarily drops and her arm moves backward. Thus, the left elbow becomes more bent like the left knee and is lifted slightly to move forward. The right foot, which points upward rapidly descends, creating a conspicuous impact.

The foot is planted with the leg perpendicular to the ground

Impact

6

carrying on the momentum

The leg and hand extend fully forward in a sliding motion. The arm swings approximately until a maximum point. The hips (the right raised higher than the left) and shoulders (the left raised higher than the right) twist slightly in opposite directions, while the areas around them and the center of the head do not move. The head remains level and exhibits an up-and-down motion, but the neck does not tilt. The left hand and leg stop in place. The nonparallel lines drawn through the shoulders and the hips cross each other. Thus, the spine is slightly flexed laterally. The area around the belly button and the middle of the forehead remains aligned to the center. If these two points swing, control over the center of gravity is lost, either from front to back or from side to side.

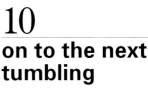

Getting off balance in a forward motion

The body slides forward.

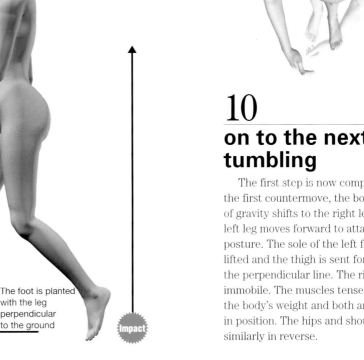

The control point for the center of gravity does not move.

The body twists

10

on to the next tumbling

The first step is now completed. As the first countermove, the body's center of gravity shifts to the right leg while the left leg moves forward to attain a posture. The sole of the left foot is also lifted and the thigh is sent forward past the perpendicular line. The right foot is immobile. The muscles tense to support the body's weight and both arms return in position. The hips and shoulders twist similarly in reverse.

12

sending the left leg forward

The action is half completed. The position is practically the flip side of the action at the very beginning, with the arms swaying lazily. The left leg is comfortably in front, with the bottom of the foot looking back and the toes slightly raised. The left side of the waist is still lower because the left leg is acting as the pivot to propel the body forward.

14

getting off balance ☜1

This is the flip side of frame 2. From here on out, let's examine some of the finer details of this action and study aspects of different walks. There are all sorts of walking styles. One striking example is the big restrictions inflicted on a person wearing high heels.

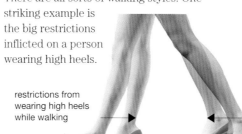

restrictions from wearing high heels while walking

16

advancing the left leg farther ☜2

This is the flip side of frame 4. Women have restrictions inflicted on their walking by not only high heels, but also the shape and length of their skirts. In this frame, the material of the skirt would press against the top of the thigh, and from there down, horizontal wrinkles would be seen.

18

carrying the momentum on

This is the flip side of frame 6. Let's check out the movements of the arms in more detail. The left arm is returning forward. The elbow is straight, and the fingers are bent back. The arm is slightly farther out than the shoulder. The right arm is swinging forward. The elbow is bent and the forearm is thrown forward and is twisting slightly inward. The palm is facing down, and the wrist is bent slightly downward. Neither hand is exerting itself and is just going with the flow—slightly rounded and spread out.

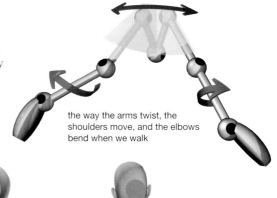

the way the arms twist, the shoulders move, and the elbows bend when we walk

20

landing and its impact

This is the flip side of frame 8. The thigh muscles on the left leg are flexing. The spine is slightly bent to the side, but the navel and the middle forehead are still in a straight line. From the side, the head looks like it is out in front. The waist is a little bit to the left and forward, and the right shoulder is slightly in front.

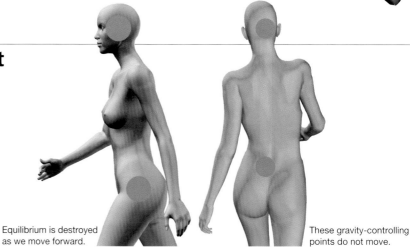

Equilibrium is destroyed as we move forward.

These gravity-controlling points do not move.

☜1 ☜2. **Restrictions on walking**

We naturally change our posture and speed when walking, depending on whether we are tired, in a hurry, etc.

● **Women's high heels**

High heels inflict a big restriction on women who wear them. These shoes narrow the length of a step because they limit the movement of the heels. We can also see the importance of the toes when we watch someone walk in high heels. The reason women can't take bigger steps when wearing high heels is that the toes are already fully extended and can't stretch anymore. The reason the feet can't stay behind longer is that the toes are bent and can't support the body's weight. It is more difficult to balance in high heels as well, which is why women take smaller steps when wearing them.

● **Women's skirts**

Tight skirts that come to just above or below the knee do not hinder a woman's walk as much when they have a slit in the side. Tight skirts enhance the bend of the back leg and keep the front leg from advancing much. It is worth the time and effort to study the construction of skirts and know how creases appear on them when a person is walking.

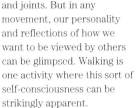
W
walking

Movements

Exploring the
Mechanisms of
Body Movements

Part 2

walking

22

on to the next tumbling

This is the flip side of frame 10. Let's investigate the head. It stays level, though it goes up and own a little because it can't stay at the same height while the legs alternately take steps. The neck is not straining, but it is ready to suddenly lean to either side if it needs to absorb a blow. That's why the head can stay still during walking, as long as there are only small movements from the shoulders down. Of course, the head is not completely immune to reactions; it moves just enough to not affect its field of vision. Note that the head is usually leaning a little bit forward.

24

again with the right leg

This is the same as the first frame. But this is not the end. Like in the other frames, the body is ready to lead into another action. This frame can link with the second frame if the walk continues.

E T C E T E R A

Walks change according to the physique and age of the person

There are all sorts of walks. They change according to people's figure, age, and personality. Fat people place more burden on their legs. To shift their body weight, they control their posture in a very different way. The arms create a big half circle as they swing along the fat spilling over the sides of the stomach. The body weight is constantly pulling forward, so the upper body leans backward. Because the range of vision around the feet is limited, movements tend to be slow.

But age is probably an even bigger factor in walking than physique. Infants who can't control their posture and children who let their actions be led by what interests them don't walk straight. Adolescents walk in a simple way because they haven't developed a self-conscious style yet. Self-consciousness sprouts in adults, and walking changes in all sorts of ways. At age 40 or so, walking feels like exercise, as it tires us. The elderly may use canes or have to muster up all their might to move.

In the first half of our lives, the reasons that our walks differ tend to be inside ourselves. In the second half, the reasons have more to do with the condition of our bodies, especially our hips and joints. But in any movement, our personality and reflections of how we want to be viewed by others can be glimpsed. Walking is one activity where this sort of self-consciousness can be strikingly apparent.

People rarely walk in a straight line and with the same feeling. Many reasons exist to make this so. From physical reasons like differences in the grade of the path to reasons driven by personal feelings, we manage all sorts of things as we walk. Make sure you don't have the footprints falling in a uniform pattern, as if they were all measured out.

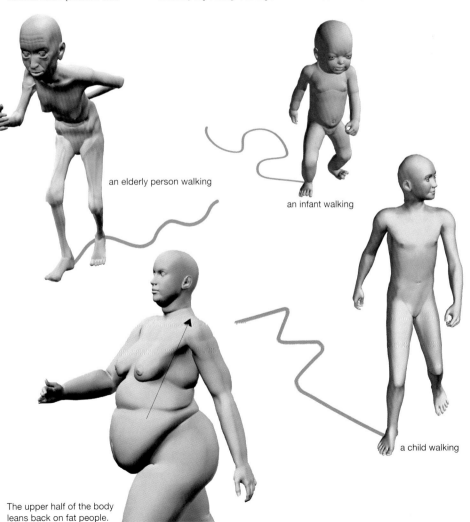

an elderly person walking

an infant walking

a child walking

The upper half of the body leans back on fat people.

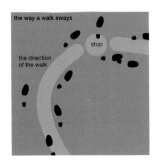

the way a walk sways

stop

the direction of the walk

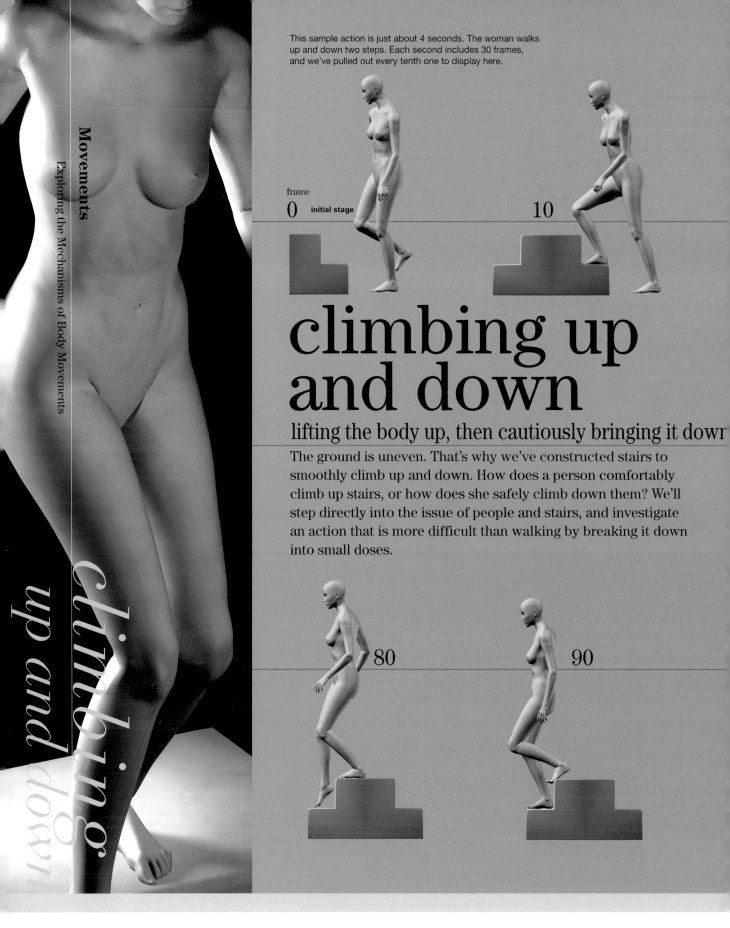

This sample action is just about 4 seconds. The woman walks up and down two steps. Each second includes 30 frames, and we've pulled out every tenth one to display here.

frame

0   initial stage

10

climbing up and down

lifting the body up, then cautiously bringing it down

The ground is uneven. That's why we've constructed stairs to smoothly climb up and down. How does a person comfortably climb up stairs, or how does she safely climb down them? We'll step directly into the issue of people and stairs, and investigate an action that is more difficult than walking by breaking it down into small doses.

80

90

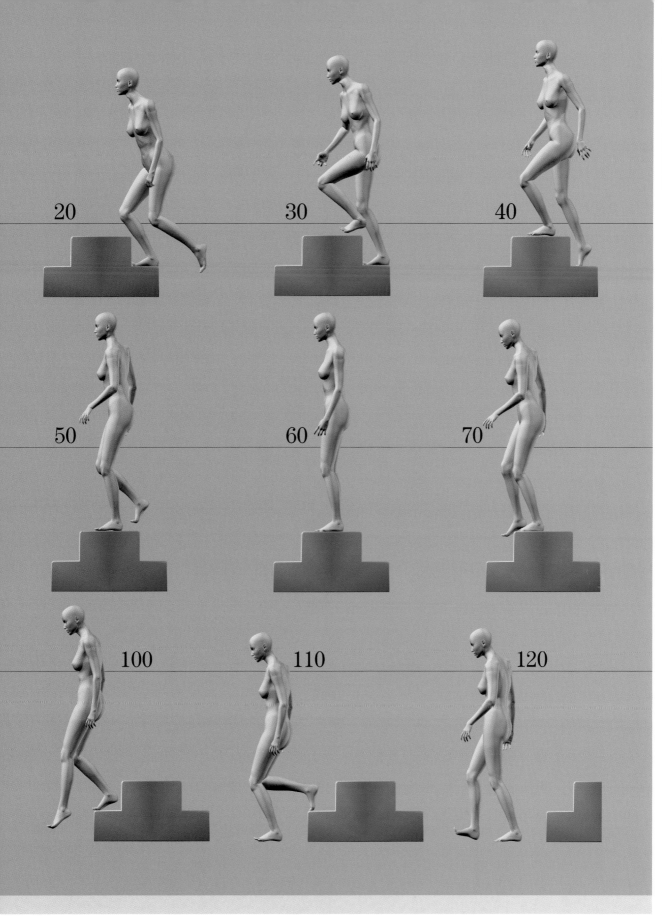

C
climbing up
and down

Movements

Exploring the
Mechanisms of
Body Movements

Part 2

climbing
up and
down

20

30

40

50

60

70

100

110

120

0
checking the steps

The steep steps used in this sample clearly show how the body changes. Most steps are not as high. For starters, as the person approaches the step, she looks at the first one. She then lifts her right foot to place it on the step. The hip is facing slightly out and toward the right.

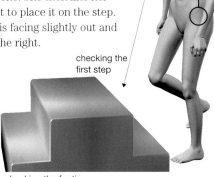

sloping back

checking the first step

checking the footing

10
onto the first step ☞1

The right foot comes down on the first step. The left arm reacts by swinging forward and up. The right leg is exerting itself. This is the moment where the body weight is shifted up. The action at this point is very similar to walking. The head is angled down a little as the eyes check the next step. The back is sloped and the arms hang at the sides.

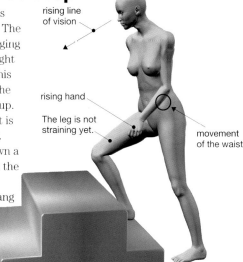

rising line of vision

rising hand

The leg is not straining yet.

movement of the waist

creating the energy to climb

20
bringing the body weight up ☞2

By using the strength of the right leg and the forward-leaning tendency of the upper body, the waist is moved. By using these reactions, the body weight shifts and the person successfully climbs the first step. The upper half of the body leans quite a bit, and the chin juts out. The hands float in the air—they have nothing useful to do at the moment. The waist and the rest of the body move to the left and forward. The right leg, which is supporting the body by itself, is bent. The body is being lifted into the air by the person's energy as well; the leg does not have to bear the full weight of the person.

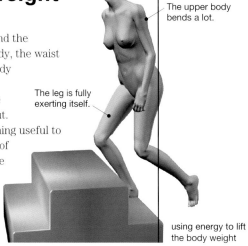

The upper body bends a lot.

The leg is fully exerting itself.

using energy to lift the body weight

30
setting off for the last step

This is just before the person steps on the top stair. The left leg has yet to touch down on the surface. In order to come to a stop on the top stair, the body begins to shut down the continuing reactions that would normally take place. Most of the body weight is shifted to the slightly bent right leg. The upper half of the body is leaning forward a tad, and the waist is still directly above the right leg. The right arm swings forward.

Opposing gravity on the way up; being pulled by it on the way down

Stairs are made so that people can climb them one at a time and descend them as well. When moving across a surface that has different grades, the body bears a greater burden than when it is just walking. When climbing, the body has to lift all its weight while opposing the pull of gravity; when descending, the body has to control its reactions as it shifts its weight lower. Going down is harder on the legs and hips as people are pulled by gravity and are careful not to fall.

People move up and down at a slower speed than when they walk. When going up, the energy needed to lift the body slows them up; when going down, the body is slowed because it is careful not to misstep. Also, the head is placed differently than when it is walking. This is a unique form of movement, and people must control their center of gravity, which is why we should pay close attention to the head's placement.

It's also necessary to understand how stairs are made. The angle of the stairs, their width, the presence of a railing or antislip material—stairs are made of many components that affect the way we move.

☞1. Onto the first step—how the parts move

The hips are still in the back, with the leg closest to the ground. The head, which was looking down, shows its intention to move ahead by sticking out. The upper half of the body leans forward. The left leg helps by staying where it is.

☞2. The line of vision when bringing the body weight up

If someone is climbing a long flight of stairs, that person can estimate the height of the next step, so he or she usually looks farther up the stairs. But if the staircase is like those short stone staircases found on walking trails, a person might not see that one of the stones is missing and step right into the hole. We check out the first step carefully, then we tend to look farther up the path.

40

arriving on the last step

The left leg exerts itself as it lands on the top step. The right leg's role is almost finished. The upper half of the body, which had been leaning forward up till now, straightens up. This is the body's way of preparing to balance while standing still—it shuts down all of the impulses to continue on. The head, which had been angled down, returns to a level position. This is the posture of someone who has just climbed a hill and finds herself on level ground. The hips are low. They have carried weight to this point. After this, they will be in a higher position.

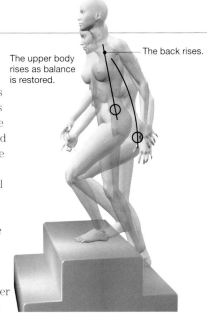

The upper body rises as balance is restored.

The back rises.

50

just before the pause

The left leg has jumped up and is stretching, bringing the right leg up in reaction. The waist is still being pulled a little, and the body weight has yet to settle. All of the weight is on the left leg, and the right leg is coming to help. Once the legs are even, the body goes into a resting posture. This is the point just before the body is still, so all of the movements are slowed down substantially.

60

on the top step

The body is standing still. The waist is stabilized. Both feet are standing evenly. Both arms are hanging down without any sign of exertion, even though they are swaying. The upper body is stretching back. The chin is slightly drawn in. The shoulders are full, and the knees are straight. Note that if you draw a straight line from the heels up, the waist would be slightly ahead of the heels. The shoulders are behind the center of the waist, and the head is in front of it. This is how the body balances itself. The action has been completed, but a person can't stay like this forever. The joints begin to bend as they prepare for the next action.

The whole body is slightly leaning.

a resting posture in the middle of the action

70

taking the first step cautiously

From here, the body descends the stairs. This is the point when the first step is made. The waist is a little lower but in the same place as before. The left leg supports the body weight, and the right leg moves cautiously forward. The right leg is extended as it searches for a spot on the lower step. Both arms help the body balance by being slightly raised and to the sides. The right arm maintains the same position as before, while the left arm moves to try to compensate for the imbalance caused by the right leg's action. The body is leaning forward, and the head is looking down. Just as when the body takes its first steps up, it is very careful when taking the first step down. The shoulders are rounded and behind the hips. From this position, it is still possible to stop descending and return to the top.

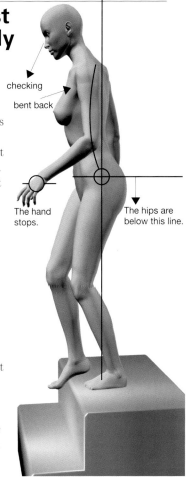

checking

bent back

The hand stops.

The hips are below this line.

descending uncertainly

80

letting the body fall ☞3

The right leg is descending to the lower step. The body remains straight as it goes down. The right foot has yet to touch the lower step, but the toes are fully extended. The left leg supports the body weight as much as possible as it tries to alleviate the impact of landing on the step.

The upper half is straight as it descends.

The lower half slides down at an angle.

The foot doesn't move.

The toes head toward the next step.

stepping down

90

onto the next step ☞4

The right leg lands on the lower step. The knee is bent to soften the reaction to the landing. The right leg is using a lot of its strength. The arms are not exerting themselves, but they are stretching down from the impact of the landing. Once they've fully extended, they sort of bounce back up. The jaw is drawn in; the eyes are looking farther ahead. The left leg begins quickly to head for the next step; it almost slides off the step.

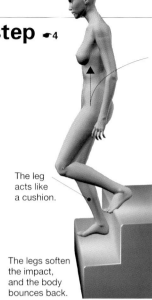

The leg acts like a cushion.

The legs soften the impact, and the body bounces back.

100

landing and its impact ☞5

The legs stretch out for the final step. This step is different from the last; the groin is stretched further. This is because the steps before this were narrow, and the leg has arrived at a wider surface to step on.

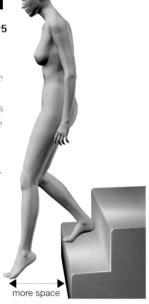

more space

☞3. **The cautious first step**

We slowly start our descent on stairs through the efforts of the left leg and the way the waist remains in a higher position. The first step is especially cautious. The rhythm and movement of actions up to now change at this point, no matter what the preceding actions were. This is an interesting point to investigate more closely.

☞4. **The impact and range of vision when the next step begins**

Basically, people don't watch each step they make. Unless there is some special reason—like dangerous material afoot—people don't just look down as they advance.

The impact of landing can be seen in the muscles and joints, but it also

can be seen in a woman's breasts or hair. The body sinks for an instant, then bounces back to normal. The movements are fast because they are falling actions.

☞5. **The last step is full of relief**

When the bottom is reached, movements are influenced by physical and mental feelings of relief

at being released from the restrictions of the stairs. This movement, free from the tension of previous movements, can be a little negligent. The body doesn't support itself as strongly, and the impact of this step can be stronger.

110
carrying the momentum on

This is the last frame of the descent. Once the ground is reached, the body is propelled forward. To control this, the legs take shorter steps. The head is forward as it checks the place the body is heading towards. The upper body leans to the right, just as the front foot does. The knee cushions the impact by bending. The waist also lowers. When the waist is in a lower position, it is proof that the leg muscles, not the bones, are supporting the body weight.

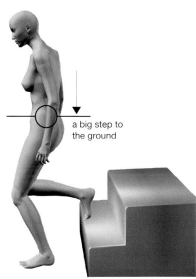

a big step to the ground

120
back to normal walking

The momentum propelling the person after the last step to the ground is controlled in one step. From here, the person can walk at a normal pace. The descent of the stairs is complete, but if the staircase was a long one, a throbbing can be felt in the thighs. It seems that we would tire more on the way up, but actually the back and legs have more of a burden to bear on the way down. To soften the blow of going down the stairs, the legs have exerted themselves quite a lot.

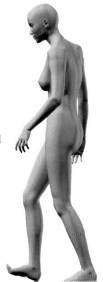
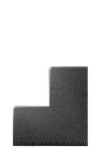

the shape and structure of stairs

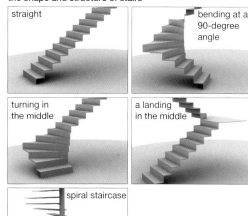

straight

bending at a 90-degree angle

turning in the middle

a landing in the middle

spiral staircase

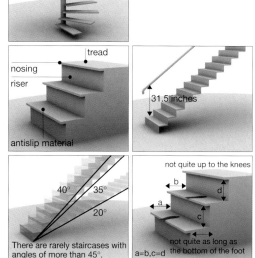

tread

nosing

riser

antislip material

31.5 inches

40° 35°

20°

There are rarely staircases with angles of more than 45°.

not quite up to the knees

b

d

a

c

not quite as long as the bottom of the foot

a=b,c=d

ETCETERA
A variety of stairs

Generally speaking, there are five types of stairs: straight stair; ones that bend at a 90-degree angle; ones that bend in the middle; ones that have a landing in the middle; and spiral staircases. The top of the stair is called the tread. The part that is perpendicular to the ground is the riser. The part of the stair that juts out a little and has antislip material on it is called the nosing. It is there to make the steps just a little bit longer. If there is a handrail, it is usually 31.5 inches above the steps. There are very few stairs sloped at 45 degrees or more, or steps shorter than 7.9 inches.

The structure of steps; the effect of shoes

When climbing the stairs, we lift a leg, advance the hips, and shift the body weight with a burst of energy. If the stair is short, the strength of the leg and the movement of the upper body is enough to advance. But the first step involves bigger movements than the other steps. From the second step, we can use the energy from the first and get into a good tempo as we climb.

In this sample, we've used two tall steps, but in reality stairs are rarely that tall. Some stairs are very difficult to climb, and this creates different body movements than those seen in this example. To capture movement on these sorts of stairs, closer study is necessary.

We also need to pay close attention to instances where the person is wearing boots that immobilize the ankles. When climbing and descending stairs, if we can't use our ankles to help our feet land properly, the body weight won't be supported and our posture becomes difficult to control. Sometimes, a person won't be able to bend her knee because the back leg can't support the body weight or the ankle of the front leg is unable to move. When we look at instances where a person is disabled in some way, it is easy to see the roles of different body parts.

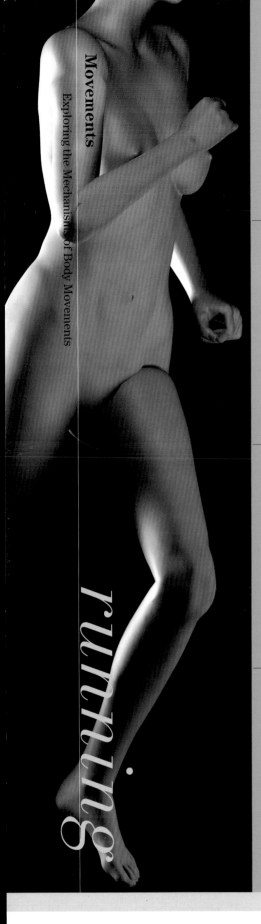

running

This sample action is of someone who has been running completely unencumbered for a while. This action shows 1 second of the running in 30 frames. We've provided almost every other frame here.

frame

0 **initial stage**

4

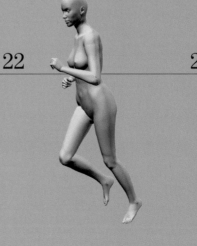

running

the continuously bounding body

Running is not an everyday activity for everyone. Perhaps that's why we can see an elegance and refined beauty in this action. It's an action where half the body is working very hard. It's also an action that isn't always done in an ideal way. In this chapter, we will explain and examine the physical side of running from the viewpoint of someone creating animation of the human body. We will also look at the basics of running and how circumstances can transform a person's run.

22

24

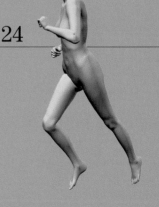

r

running

Movements

Exploring the
Mechanisms of
Body Movements

Part 2

running

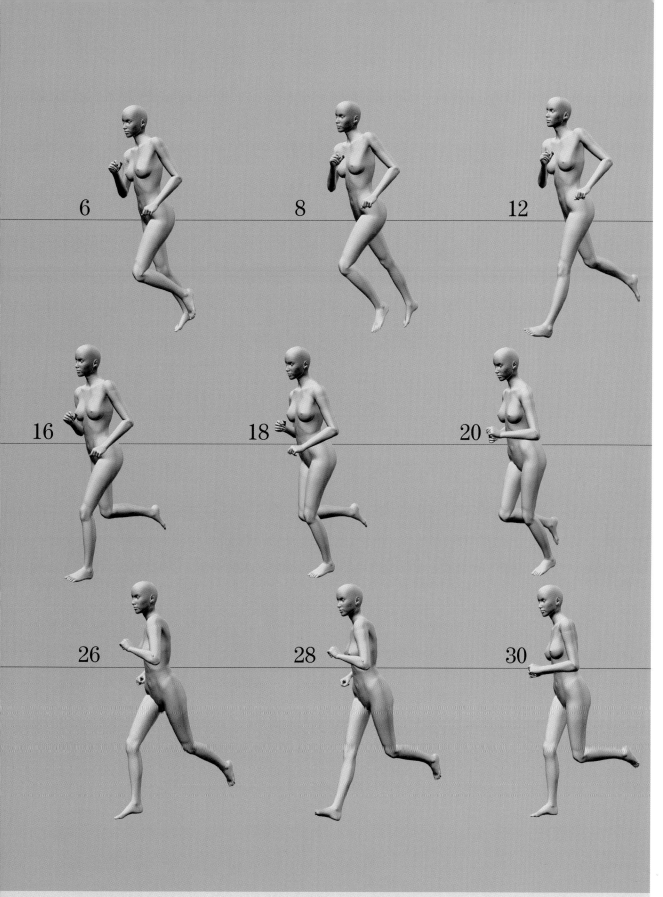

6

8

12

16

18

20

26

28

30

0

impact of landing ☜1

This frame captures the moment that the right foot lands on the ground. The right leg is using a lot of strength, and most of the impact of landing is absorbed by the knee. Even so, the whole body shakes from the impact. Because the leg is coming up again soon, the hands don't drop.

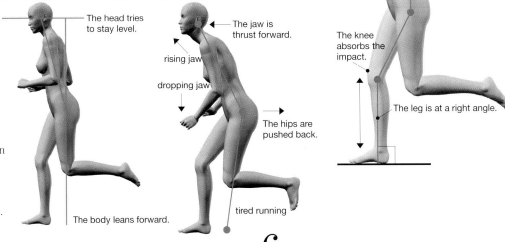

The head tries to stay level.

The jaw is thrust forward.

rising jaw

dropping jaw

The hips are pushed back.

tired running

The knee absorbs the impact.

The leg is at a right angle.

The body leans forward.

4

body sinking down

Once the right foot has landed, the body is propelled forward, pushing that foot to the rear. When you look at someone running, this is the lowest point. The right arm stretches out before the rising left thigh, which is also heading forward. When the right foot hit the ground, the whole body felt the impact, and women's breasts sag for a moment. The jaw also drops a little, although the head is trying as much as possible to stay level. The shoulders are sunken. From here, the whole body begins to rise.

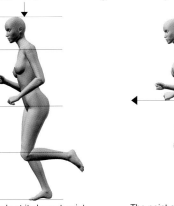

the body at its lowest point

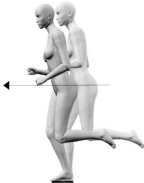

The point of contact doesn't move.

6

body bouncing up

In one burst, the body tries to rise diagonally. The shoulders shrug as they rise. The right arm moves closer to the chest to maintain balance. The fist rises first; the shoulders follow. Remember that the fist's rise is especially fast. It is driven to rise; the hand uses its strength to stabilize the body. The underarms are firmly shut. The left thigh is exerting itself and is just about to kick up.

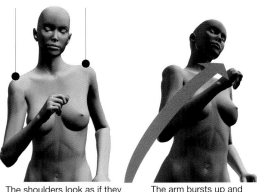

The shoulders look as if they were suspended from above.

The arm bursts up and near the body.

Three telling characteristics of human movement

The three most important characteristics of human movement for animators are consciousness, weight, and habit. Consciousness refers to what a person is interested in and where he or she is looking. It also refers to separating consciously made movements from subconscious ones. Weight affects possible reactions to movements, including preparatory and successive actions. Also the animator must consider how weight is supported and balance is maintained. The final characteristic, habit, is about individual differences. Movements can convey differences in gender, age, and race, and bring out the character and circumstances surrounding the individual. These three categories are prime subjects for further investigation and must be enhanced in the depiction of the human body.

Another important and fundamental topic is the understanding of the human structure. We move around without much thought of how our bodies work, but actually human bodies have shapes formed by movements and movements formed by shapes. To use a very simple example, because we have two legs, we necessarily tend to walk in a certain way. However, we weren't given two legs for the purpose of walking. Once we deeply understand the human structure, we can probe beyond the general circumstances and imagine more movements.

☜1. The impact of landing— posture and body weight

Look at the picture in frame 0. This is one picture taken of a woman in the middle of a run. The upper body is leaning forward a bit, the head is raised, and the eyes are staring ahead. Usually, the jaw is slightly drawn in when running. When a runner gets tired, the jaw begins to jut out. This movement forces the shoulders out, slopes the back, and pushes the hips back.

r
running

Movements

Exploring the
Mechanisms of
Body Movements

Part 2

running

8

body stretching up

The left leg has risen. This gives birth to the biggest reaction. To run faster, the thigh rises higher. The waist and shoulders twist deeply to retain the equilibrium on either side. The right foot is pushing off the ground at this moment. The body is leaning forward and pushing up. The right arm stays up. At this point, a woman's breasts would be flung to the left and up. The upper body also twists this way. The left arm is bent at the elbow as it pushes back. This is the point where the body is at its highest, as the shoulders, arms, and legs all push the body up. From here, the body sinks.

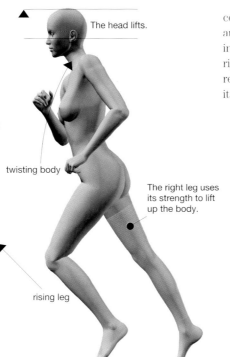

The head lifts.

twisting body

rising leg

The right leg uses its strength to lift up the body.

12

body falling

The body begins to fall. The arms and shoulders continue their movements. The elbows, shoulders, and chin stay raised. The left leg stretches out quickly in an effort to land as far forward as possible. The right leg has left the ground and is released to the rear. The body looks as if energy is being drawn from it as it falls forward and prepares for the next impact.

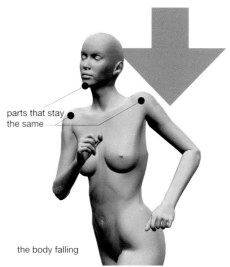

parts that stay the same

the body falling

16

landing and proceeding

Now the left leg lands. The left leg below the knee is at a right angle with the ground as the knee and thigh try to absorb the impact of landing. The body continues to sink. The image here—stable and strong—is different from before.

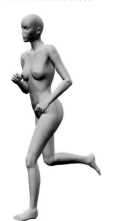

18

preparing for the next jump

The body continues to sink. The left leg is preparing to push the body upward. The right hand also subconsciously comes back. The left hand begins to lift the upper body. It is trying to keep the right leg from turning the body around

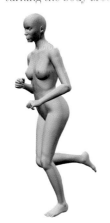

20

carrying the momentum on

The body has sunk again. The left leg is bent, and the whole body is low. A low stance indicates that the body is carrying a burden. Because the person leaped forward, she must sink. This sort of compensation—to move faster, a person has to bear more of a burden—is what tires a runner out over the long haul

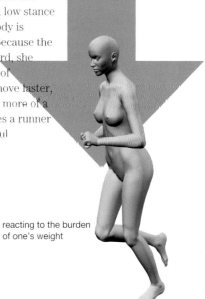

reacting to the burden of one's weight

22
body stretching up

The body begins to stretch out again. The left leg uses a lot of strength to propel the body upward. The left arm has also risen. The right leg is already coming to the front. At this point, the toes on the left foot are tightly gripping the ground. For comparison, the toes on the right foot are just being carried along. The left arm is making conscious actions, and the upper part is exerting itself.

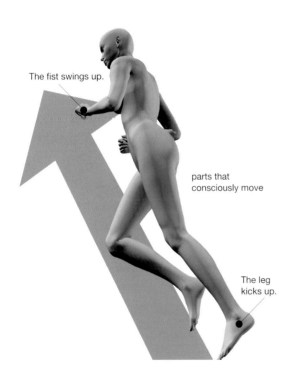

The fist swings up.

parts that consciously move

The leg kicks up.

24
legs at full stride

The left and right legs are at their most distant. The left leg is once again off the floor and kicking backwards. The left shoulder is pushing forward. The head tries to stay level but it still moves quite a bit because the upper body is twisting a lot during the run.

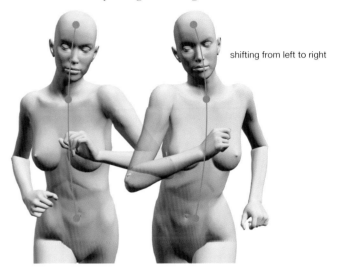

shifting from left to right

26
just before landing

The right leg is extended and near the landing point. Note that the shoulder will swing out more than usual if the runner is carrying something or can't swing her arms for other reasons.

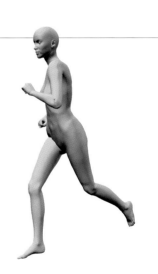

Running without swinging the arms

If we can't swing our arms when running, we become unstable and we can't lift our legs as high, which reduces our speed. To run fast without using our arms, we may try to make our strides longer or make the legs go back and forth faster. With the former, the body will rise more, making the impact upon landing stronger, but the movements are slow, giving the body stability. With the latter, the body minces forward in an agitated fashion. While the impact of one step is slight, the body constantly swings wildly, creating instability.

Out-of-control, wasteful actions

When we tire, our movements become confused. We really don't want to use our muscles any more, so we try to make our landings softer, and our arms swing feebly. As we mentioned in the first frame of this sample, the back is rounded and the chin sticks out. The feet get louder as they land, and the impact vibrates through the whole body. The hips and shoulders slip further to the left and right or forward and back, and the whole movement slows down.

Turning awareness on and off

Running is an action that clearly shows when awareness or consciousness is "on" and "off." The arm comes forward with intention, which means it's "on." But it's "off" when it swings back. The legs are "on" when they thrust forward or when they kick off the ground, but they're "off" when they extend backward. This doesn't mean that runners consciously think that it's

r running

Movements

Exploring the
Mechanisms of
Body Movements

Part 2

running

28
body stretching up

The legs stretch out again. The extended leg heads toward the ground, and the knee bends just before impact. The median line of the body, which runs through the center and connects the central spots in the shoulders and the waist, is a very important tool in understanding the upper body at all times, not just when running. If these two spots start to slip out of line, the head or legs will always try to compensate. If they slip too far, the runner could fall over. Note that with any movement, the connection of these two spots and a third spot where the feet hit the ground dictates the movements of the arms and legs. The median line is especially important for a person standing straight.

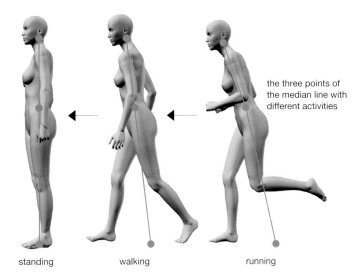

the three points of
the median line with
different activities

standing   walking   running

30
at the end of a full cycle

This frame is the same as the first one. The impact of landing is running through parts of the body. This frame can be seen as both the moment of the biggest vibration and the stillest moment. Even as the run slows, it is still a run and not a walk. The difference is in how the leg kicks up from the ground during a run. In fact, this is the most distinct characteristic of running. It doesn't even occur in fast walking.

Now you can begin to see when the shoulders, hands, hips, and other parts are at their highest and lowest points, and when an action is intended or involuntary.

the successive movements of different body parts

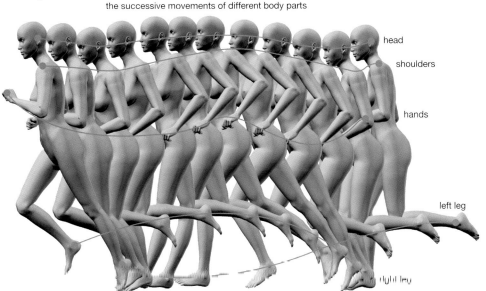

head

shoulders

hands

left leg

right leg

time to swing their arms. Running is something we have become used to, so we run in a manner that is most comfortable to us.

Any action has levels of mastery. Running is not something we do every day, which is why differences emerge. An extremely different form of running would be skipping. People don't run that much. Running puts a big burden on the body, which makes us reluctant to run again. If it's not for exercise or training or some other voluntary reason, most people run as if they have no choice.

Usually, people don't run in a straight line; they may turn as they run. They tend to form a bit of an arc because of their momentum. When turning a corner, it's dangerous to just check for oncoming traffic at the last second, so runners slow down by leaning their upper body back and putting a stop to their reactions.

Also, running for a long time and running for short bursts are clearly different. It's the difference between pacing yourself and going for speed. When running, toes play a significant role in lifting up the body. So shoes—with hard, unbendable soles—seen on hard leather shoes, for example–are not suitable for running. They don't allow the toes to bend on the ground to lift the body. For the same reason, high heels and loose sandals are also unsuitable. Boots and high-top shoes restrict the ankles, and backpacks put a burden on the body, making it increase its reactions to compensate. We need to consider all of these points and more once we have understood the basics of running.

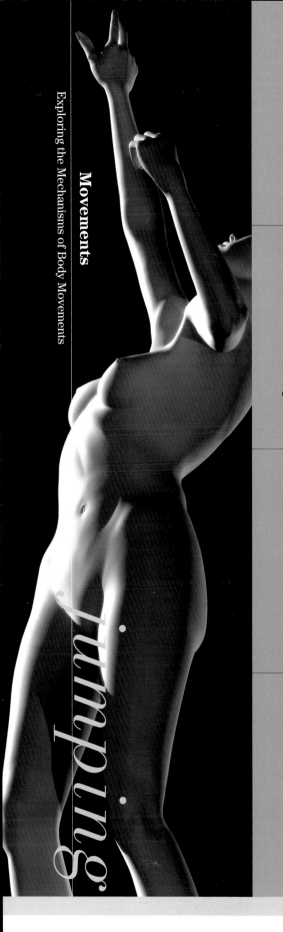

This sample action includes three continuous actions—visualizing the feat, jumping, and landing—that take about 3 seconds to complete. With 1 second taking 30 frames, this action has 85 frames. We've chosen almost every seventh one to display here.

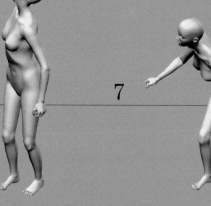

frame

0   initial stage

7

jumping

the shape of consciousness

Now for jumping. We'll focus on the mainly conscious decisions made here, from the difficult task of controlling the body weight to the building of the strength needed to make a jump. With any action, there is first an intention and then the action trying to fulfill that intention. In the continuous act of jumping, what intentions are at work as movement occurs? Let's take a look at a vertical jump to learn.

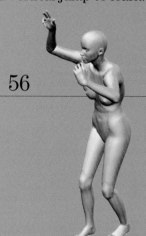

56

63

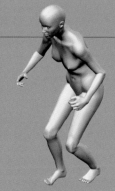

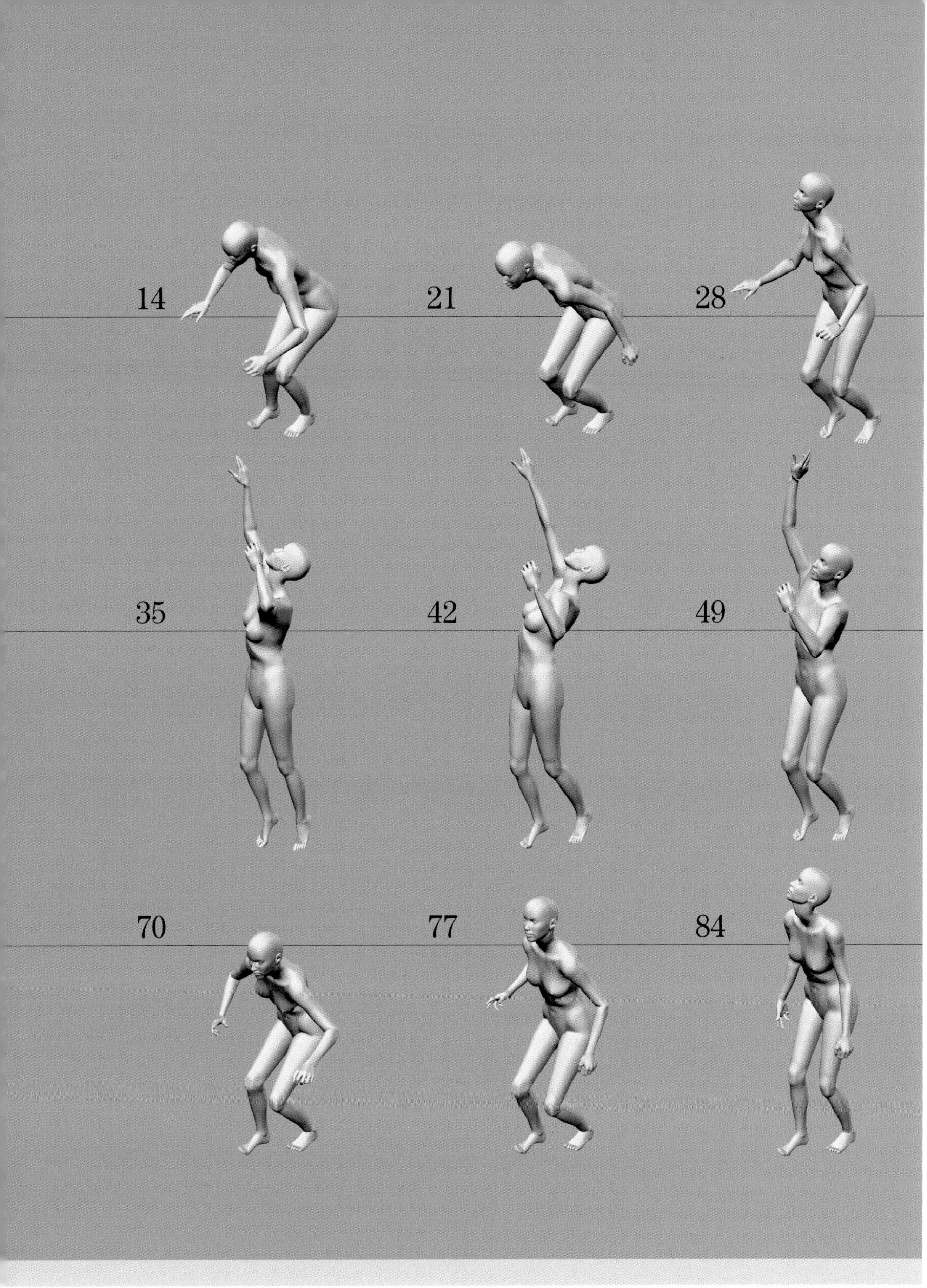

j

jumping

Movements

Exploring the
Mechanisms of
Body Movements

Part 2

jumping

14

21

28

35

42

49

70

77

84

0

visualizing the feat ↜1

First the hand stretches up toward the target and confirms where to aim. This is more than just checking on the target—it's like rehearsing the jump. Often the person will stretch her arm a lot as she marks off a spot as high as she thinks she can jump.

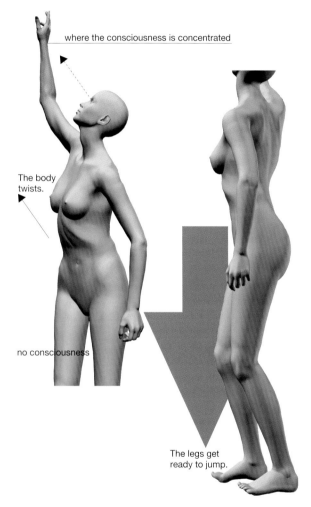

where the consciousness is concentrated

The body twists.

no consciousness

The legs get ready to jump.

7

bending the body forward

The body gets ready to react. The upper body sinks and leans forward. Now, as the body sinks, the consciousness switches from the target above to the space in front of the knees. We treat our face as the most important body part, and consciousness follows it. This rule is not limited to the conscious moving of arms and legs. In this frame, the head is still slightly raised as it just begins to change its focus. The hips are pushed back, and the chin is thrust out. The arms aren't really being controlled, as they're not consciously involved in the action. They are hanging and slightly separated from the body as they help it keep its equilibrium and wait for the chance to jump.

upper body leans

line of sight changing

hips sink

14

sinking the body down

The body begins to sink. The shoulders drop, and the back rounds. The face looks down; the eyes gaze out at nothing in particular. The hips are pushed farther out and down, and the back becomes even rounder. The head and hips work at keeping the body balanced. The arms, until now unaware of the action, begin to exert themselves to prepare for being held aloft by rising a little and extending outward. This is the moment where the action switches from sinking to reacting.

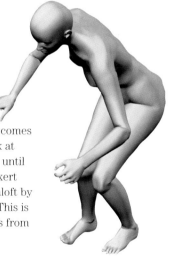

The shape of conscious movements

There are many ways to jump, but we selected the vertical jump because the conscious transitions that take place are striking. The vertical jump is used in physical examinations to measure a person's vertical spring. The reaching hand will touch a line on the blackboard that shows the height of the jump.

Basically, the person doing the jumping will aim for a spot as high as physically possible. Not much

thought is given to what happens after the jump, which is why people often fall after an attempt. If a person is planning to make a high jump, her actions before the jump become bigger. The eyes are fixated on the target, and the fingertips pulsate with awareness. The whole body sinks deeply, and the arms swing a lot to build toward a big reaction. While these actions are taking place, the consciousness is concentrated on jumping high, and the head is eyeing the target. At the same time, the body

is subconsciously working to make the jump as high as possible. The intensity of the action to come is revealed by the line of sight, which is steadily focused on the target.

People rarely jump without a purpose. This is true about other actions as well.

↜1. Visualizing the feat—the movement and consciousness of different parts

The head is pointed at the target. The person is coming to grips with

the distance from her position to the target, and though she is not jumping, she is reaching toward the target to confirm it. Although the arm is stretching to its limit, the legs are not assisting by stretching out the back. To prepare for the jump, the knees bend slightly, creating a flexible posture, but this is done subconsciously. The upper body is twisting up and to the right. The right arm is in control, but the left arm is just hanging there.

j
jumping

Movements

Exploring the
Mechanisms of
Body Movements

Part 2

jumping

21
holding the energy

Both arms swing back forcefully in the moment before the jump. The upper part of the body sinks farther, the knees bend, and the chin juts out. The full weight of the body is being borne by the legs, and the tension is visible. The arms aren't really exerting themselves, but they are preparing to form a big arc. Up until this point, the hips have taken the initiative while sinking lower.

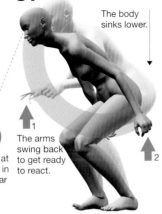

The body sinks lower.

line of sight
looking at nothing in particular

The arms swing back to get ready to react.

28
stretching up

The body is released like a spring; it begins to stretch upward. The hips are up and a little forward. Both arms are making a big arc as they swing, lifting the upper half of the body. The eyes immediately search for the target, quickly darting to the upper right. While the upper half of the body bends back, it brings the waist along. The arms help here by rotating as the body prepares to leap. From this point, the legs strengthen and provide the final kick. This moment contains very fast movement.

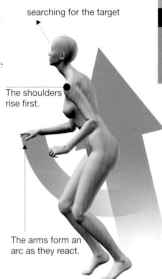

searching for the target

The shoulders rise first.

The arms form an arc as they react.

35
jumping ☞2

Both legs stretch out strongly and the body flies into the air. The strength of the legs is gone in an instant, and they now move in a slightly lax way. The whole body extends, and the shoulders rise. In this instant, the body parts that are under control and those that aren't are strikingly apparent.

Consciousness is concentrated here.

staring at the target

The upper half of the body twists as the legs rise.

not conscious of the action

42
moment of accomplishment and release

The fingers hit the target for an instant. After hitting it, the right hand heads to the back. This action was created by fully extending the upper half of the body and doing a half turn with the arms, and now that the goal has been reached, the upper body bends back. But the head is still fixed on a spot beyond where the right hand was reaching. The body is pulled by gravity and is forced to descend. The left shoulder begins to slump a little. The legs look as if they are writhing as they are released to the rear. Like the instant when a marathon runner crosses the finish line, the controlled movements give way to more relaxed movements. At this instant, the consciousness moves from the right arm to controlling the posture as the body plans its safe landing. The connection between the head and the right hand is cut. The conscious control exercised in landing is not purposeful, but instinctive.

The hand arcs up to hit the target.

The body bends back in reaction.

The feet are lax.

☞2. Jumping—the shape of movements, consciousness in different parts

The eyes have been staring at the target. The right arm extends toward it. It's not necessary for the left arm to do anything, so it just sort of hangs in the air. But note that the left shoulder is firmly raised.

Consciousness is solely focused on the right fingertips, not on the rest of the arm or the legs. The whole body is bent, putting the hips slightly in the front. The body is focused on trying to reach as high as possible with the right hand. This is a sudden burst of activity, so disorder can be seen in several places. But the head is still firmly pointing at the target, and the right hand is replying to it.

49
body falling

The body begins to descend. It reflexively prepares for the descent by bending and going into a defensive posture. At this point, the action is probably not going exactly as envisioned. Body parts are probably not doing exactly what they are supposed to, and control is only partial. The person is falling backward a bit. She is pushing her hips back and pulling them forward to retain balance. The legs are already bent and bracing to absorb the impact of the landing. From the front, the person's median line seems to be leaning to the left, making it hard to control the body weight.

The shoulder is still up during the descent.

Consciousness moves to the ground.

The hand moves from the wall.

The shoulder leans to the outside and is difficult to control.

no consciousness

As the body drops, it loses control.

56
awkward landing

This is the moment when the feet touch the ground. The legs are tense, but they are in control enough to move softly. Both shoulders are raised, and both arms follow. The back is quite rounded, but it is not leaning forward that much. The consciousness is focused on controlling the posture. The eyes aren't looking at anything in particular; they are cast downward. The head rises as the body tries to straighten the median line and restore balance. The body's flesh slumps as a reaction to the descent. The feet are on the ground, but they don't have steady footing yet.

63
absorbing the impact

The footing is eventually established. The head is still raised. The shoulders are followed by the arms on a gentle descent toward the waist. The waist is also sinking. Both thighs are tense as they brace the body.

The head is vertical.

The balance tilts to the left.

70
insecure balance

It looks as though the person has landed skillfully, but coming back from a big slip is not as easy as it looks. The balance is still a little off. To restore the balance of the upper half of the body, which was tilting to the left, the waist quickly jerks to the right and the left arm swings out. The hips and arms are releasing the impact that the legs could not absorb. The head is raised and defending its position. Once the person has taken a little time to see if she is really stable, she'll stand.

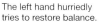

The left hand hurriedly tries to restore balance.

j
jumping

Movements

Exploring the
Mechanisms of
Body Movements

Part 2

jumping

77
back into a standing position

After balance has been restored, the body rises again. The head looks as if it is suspended from above as it stays fixed while the waist slowly rises. After the arms have finished their work, they keep balance while going slack and dropping to the sides. The hips, having been pushed to the back, now move up and forward. The body is consciously working toward a final goal of standing straight. Compare this to the instinctive shape of frame 63, and it is clear that the body intends to control its movements here.

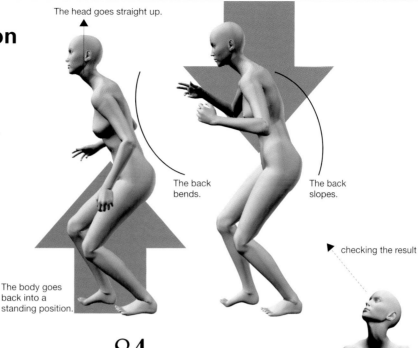

The head goes straight up.

The back bends.

The back slopes.

The body goes back into a standing position.

checking the result

ETCETERA
The processes of checking

1. sight
2. touch
3. smell
4. taste (perhaps the final goal all along?)

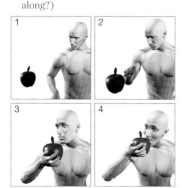

84
finishing up

All that's left from here is to straighten up the back and stretch out the knees. At this point, the head looks up and to the right for the target, searching for the mark made by the hand. The two actions—the subconscious standing and the conscious looking at the target—are done together. Of course, the person's focus is pulled toward the result of the jump.

not much awareness in these parts

checking the result while standing

The conscious and subconscious movement of body parts

Expressions of the ego are mostly depicted in actions. Even the desire to do nothing expresses an intention. Actions like sleeping on a sofa or lying on the floor with one's arms and legs spread out give us an inkling of the person. When depicting a living person with a will to do something, we must aim to show the person making a move that clearly indicates what that person wants to do.

Choosing something to do is the next step after identifying a person's interest. To know what someone's interests are, we have to study the person going through the processes of checking.

The man in the diagram (above left) goes through the processes of inspecting an apple. First, the apple is checked by the eyes. This is something we do even if we're not that interested in the apple. Next, the man touches the apple. Then, he reflexively smells it to see if it can be

eaten. If the object is the right size to put in the mouth, that is the next step. Of course, if he had inspected the apple before, he would cut out some of the steps. But be aware that all actions go through this process.

To understand the action of jumping, remember that the jumper changed her focus in several places. When checking the target, she then worked to see if she could jump even a little higher. Then she tried to touch that higher spot. After that, the jumper prepared to land. Lastly, she

checked the target again. At each stage, be clear about which parts are moving subconsciously and which parts are ignored. By paying attention to these points, you'll be able to see how a body begins to slip when trying to maintain its equilibrium. Was it a slip that came from not being able to control the body, or something caused by a part that wasn't being controlled? It's important to know what causes the swaying in the body.

The sample action is of a right-handed man trying to throw as well as possible. The action takes about 4 seconds. At 30 frames a second, that's 120 frames. We've pulled out the most distinctive ones and explained them.

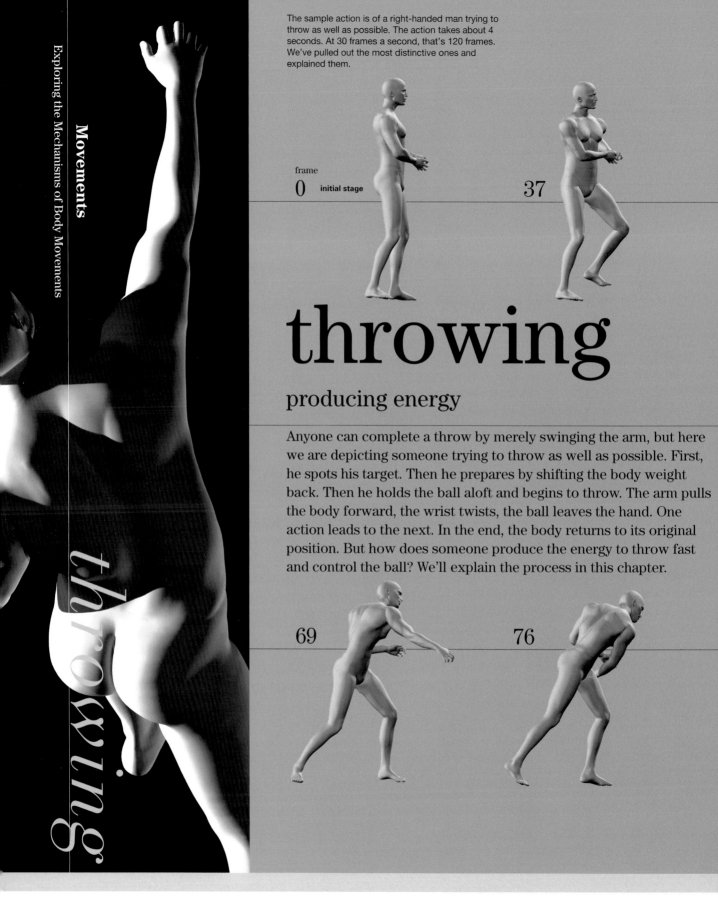

frame

0   **initial stage**

37

throwing

producing energy

Anyone can complete a throw by merely swinging the arm, but here we are depicting someone trying to throw as well as possible. First, he spots his target. Then he prepares by shifting the body weight back. Then he holds the ball aloft and begins to throw. The arm pulls the body forward, the wrist twists, the ball leaves the hand. One action leads to the next. In the end, the body returns to its original position. But how does someone produce the energy to throw fast and control the ball? We'll explain the process in this chapter.

69

76

t

throwing

Movements

Exploring the
Mechanisms of
Body Movements

Part 2

throwing

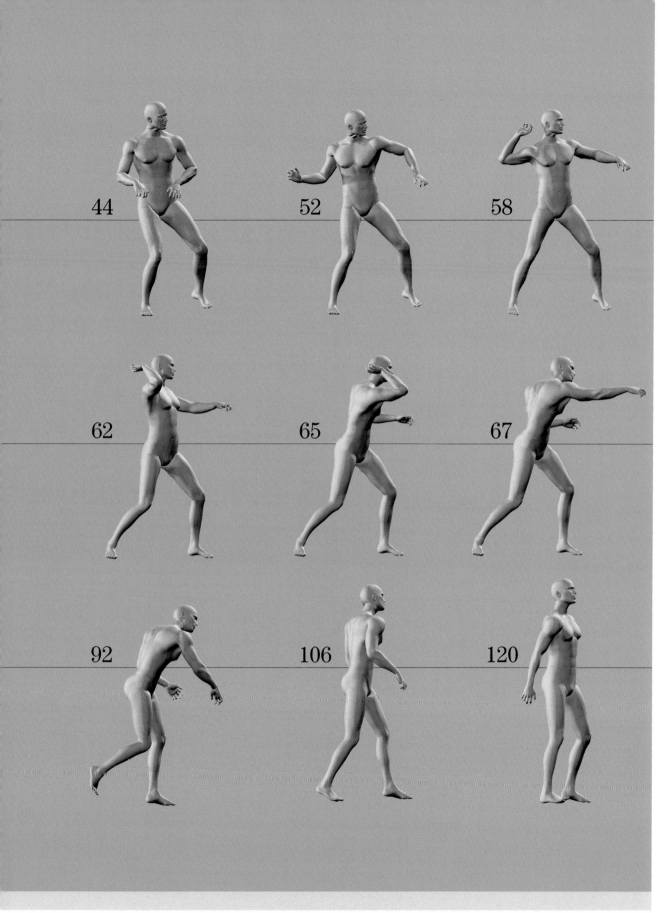

44 52 58

62 65 67

92 106 120

0

checking the target ☞1

First, the person checks the target. The upper body stands tall; the body is stable. The right leg steps back just a bit as if in preparation for the throw. The head juts forward a bit in a relaxed manner. The right hand, which holds the ball, and the left hand are held around the stomach. The shoulders are relaxed.

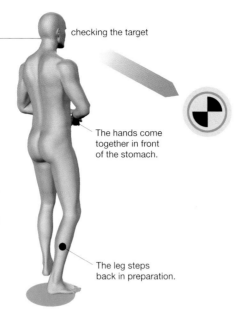

checking the target

The hands come together in front of the stomach.

The leg steps back in preparation.

37

shifting the body weight to the rear

Now he's ready to throw. The upper body turns to the right and faces the target at an angle. The left leg leaves the ground. The right arm rears back. The hips move back a little while the shoulders drop back. The head is staring at the target.

scrutinizing the target

body drops back

body rotates

right leg forms an axis and rotates

left leg lifts

44

body facing sideways ☞2

The body rotates from the hips. The chest is facing right. The upper half is back and leaning a little to the side. The right arm is bringing the ball back farther. People who haven't learned to throw well don't move sideways. They just lower their hips, push back their buttocks, and jut out their chin.

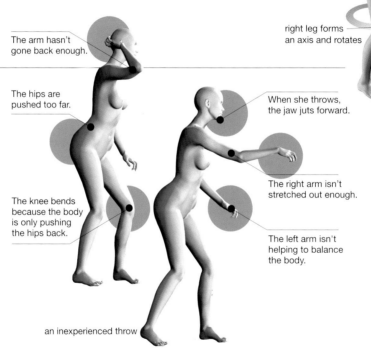

The arm hasn't gone back enough.

The hips are pushed too far.

The knee bends because the body is only pushing the hips back.

an inexperienced throw

When she throws, the jaw juts forward.

The right arm isn't stretched out enough.

The left arm isn't helping to balance the body.

☞1. Throwing strength: creating the energy for each part to rotate

Throwing is something that starts early in infancy. But with infants, there is no target and the action is reduced to grabbing and throwing. It's just grabbing and impulsively releasing things. Aiming at targets, gauging distances, and controlling strength and speed all come later. Throwing becomes a more refined activity once a person has an opportunity to play catch. Girls often don't have the chance to learn to play catch and become familiar with balls. When they grow up, they usually don't know how to throw very well. Most people try to throw the ball as fast and as accurately as they can. To do this, the linkage of different parts is indispensable, but the rotation of all of those parts is what mainly helps speed up the moment where the ball leaves the hand. The thrower is completely focused on the target.

☞2. The shape of movement in several parts

There is no consciousness in the left arm. The left hand stays in place and then moves with the body to the right. The left leg is lifted up more. All of the weight is on the right leg, which serves as an axis. The head looks at the target, but the body is turned sideways and leaning back. The chin is raised and also tilting toward the back. The back is slightly hunched, and the hips are fixed in place. The weight is shifted back until it seems as if the person might tip over.

☞3. Holding the ball aloft

The left leg, hip, and shoulder move forward in that order. The left leg has risen. Now it is extended and stretched forward. The right arm is still holding the ball aloft. The wrist begins to bend to the outside. The left arm, just hanging there until now, swings quickly to the front in reaction to the rotation.

t
throwing

Movements

Exploring the
Mechanisms of
Body Movements

Part 2

throwing

52
compound movements ☞3

From here the body moves forward in one stride. The consciousness is concentrated in the right arm as it passes energy through to the ball. The forearm from

the elbow continues to move. This is a moment when all sorts of parts are rotating in a complicated manner. However, the head stays focused straight ahead.

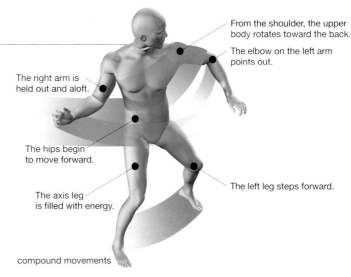

From the shoulder, the upper body rotates toward the back.

The elbow on the left arm points out.

The right arm is held out and aloft.

The hips begin to move forward.

The left leg steps forward.

The axis leg is filled with energy.

compound movements

58
first axis of rotation ☞4

The body creates energy in an instant by compounding the rotation of different parts. The right arm lines up to throw the

ball as straight as possible. The first and biggest rotation begins now—the rotation of the whole body.

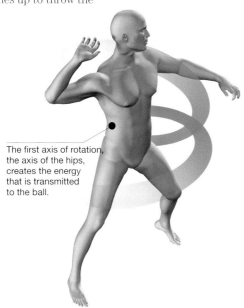

The first axis of rotation, the axis of the hips, creates the energy that is transmitted to the ball.

62
second axis of rotation ☞5

The rotation of the hips continues. The upper body proceeds to turn until facing forward. The whole body turns in an arc at this instant. It's like a taut bow about to shoot an arrow.

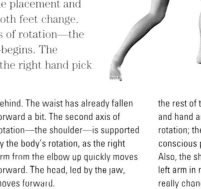

from rotation of the body to rotation of the shoulder

65
third axis of rotation ☞6

Once the hips are set in place, the rotations continue. At the same time, the placement and direction of both feet change. The third axis of rotation—the right elbow—begins. The fingertips on the right hand pick up speed.

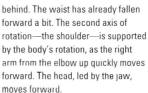

The third rotation is in the right elbow.

advancing forward

energy in the left leg

☞**4. First axis of rotation**

The body weight has been lowered sufficiently and placed on the hips. As the body moves forward, the waist rotates to the left. The legs are far apart, forcing the body to work at maintaining its equilibrium while these two big actions take place. The left leg has touched the ground. The right leg, which was bearing the body's weight, begins to transfer it to the left. At this moment, the parts that are moving the least are the hips. The right arm is finally finished holding

the ball aloft; the arm is just behind the head with the wrist bent outward. The left arm has also stretched outward as much as it can. The chest is wide open, and the body from the waist up is beginning a big rotation to the left.

☞**5. Second axis of rotation**

The right elbow has been pushed down in the rotation, but the right hand is where it was before. The chest is pushed out, the right leg is exerting itself, and the left leg is left

behind. The waist has already fallen forward a bit. The second axis of rotation—the shoulder—is supported by the body's rotation, as the right arm from the elbow up quickly moves forward. The head, led by the jaw, moves forward.

☞**6. Third axis of rotation**

The upper part of the body, which had turned back, comes bursting forward in one fluid movement. The right shoulder is raised and near the ear. The right elbow is out in front of

the rest of the body. The left elbow and hand are pulled down by the rotation; they are basically not a conscious part of the movement now. Also, the shape and placement of the left arm in relation to the body do not really change. The left arm stays close to the body and waits to see if it needs to adapt to different circumstances.

67
fourth axis of rotation and release of the ball

This is the moment when the ball leaves the hand. The body is continuing the movement of the last frame, and now the right arm is fully extended. The final axis of rotation—the wrist—gives a last spurt of energy to the ball as it is released. If you don't add any unique traits to the rotating wrist, the path of the right arm from frame 62 or so to this frame is almost completely straight. This results in the ball flying straight.

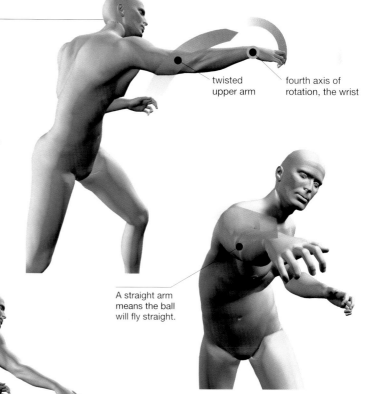

twisted
upper arm

fourth axis of
rotation, the wrist

A straight arm
means the ball
will fly straight.

69
following the ball with the eyes

The body sinks farther as it reacts. With its role done, the right arm stays fully extended as it falls in a diagonal line to the left. The head follows the ball. The right shoulder is pushed out; it leans to the right as it rotates left.

76
body curled and sunk down

The right arm swings until it hits the body. The body sinks farther. The hips are pulled forward. The head is still raised to keep the body from falling forward, but it is no longer staring straight ahead. Athletes will still be firmly focused on the ball at this point, but for an amateur who has just thrown as hard as he can, the body's rotating will lead to a rotation of the neck. All of the body's weight is on the left leg; the right leg is practically using no energy and is on the ground or almost touching it.

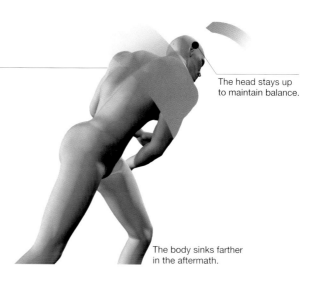

The head stays up
to maintain balance.

The body sinks farther
in the aftermath.

t
throwing

Movements

Exploring the
Mechanisms of
Body Movements

Part 2

throwing

92
releasing the momentum

The left leg extends and lifts the body, allowing the energy accumulated there to escape. The right arm counters this by rotating in a big motion to the right and back to the place it started. The right leg continues the movement it was making before. The upper body follows the right arm as it starts to rotate in the opposite direction. At this point, the person tries to raise his head. The base of the jaw and the neck move back and up, forcing the jaw to jut out a little.

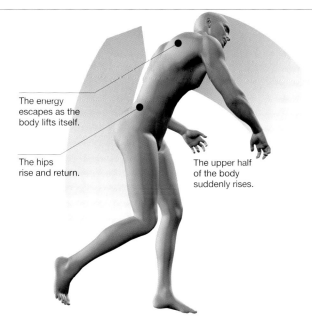

The energy escapes as the body lifts itself.

The hips rise and return.

The upper half of the body suddenly rises.

106
back to the starting posture

The right foot hits the ground. The left leg is stretched, and the hips are almost back to where they started. The back is also stretched, but it has to come back from the big movement made by the right shoulder, so that shoulder is still a little higher than normal. The right arm, from the elbow down, swings back. The neck, which had been pushed back and up by the head, keeps rising; the chin follows. Overall, it's one moment in a quick action.

120
checking the result

The person slowly checks whether his throw hit the target. As the body returns to the starting position, the right arm swings even farther back.

ETCETERA
A fluid movement using no strength

Actually, there are moments when the legs can no longer keep the body stable. It happens in the middle of the action; the body loses its balance and falls. In this case, make sure to thoroughly check the movements before and after the fall.

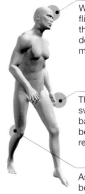

While checking the flight of the ball, the head is pulled down by the body's movements.

The arm isn't swinging to keep balance; it swings because of the reverberations.

As the body becomes unbalanced, the back leg doesn't land right.

Throwing is an action related to objects

Throwing is an action that makes all sorts of body parts mobilize together for the sake of seeing just how skillfully a ball can be released. Awareness is focused on the ball. The body combines several rotations to give the ball speed and control its flight.

As you know, the wrist itself cannot rotate. Two bones in the upper arm twist, rotating the area from the wrist on. The upper arm shows signs of the muscles twisting at this point. As the body rotates, the arm extends forward in one fluid movement and throws the ball. In this instant, rotations build one on top of the other until they peak in the fingertips by creating speed for the ball. The fingertips even fill with blood.

Remember that the job of supporting the body shifts from the right leg to the left. The lower half puts the ultimate priority on stabilizing the body. That job is shared by both legs. People can't fly, so the body's weight must always be supported somehow. If this part of the depiction is done crudely, the movements will look light and unrealistic. Also remember that

throwing has a fluidity that slowly builds to an agile throw and a leisurely return.

Where is energy being expended? How will it affect the ball? Some parts are following through with movements they originally had no intention of making. It's important to understand this fluid movement that doesn't use strength.

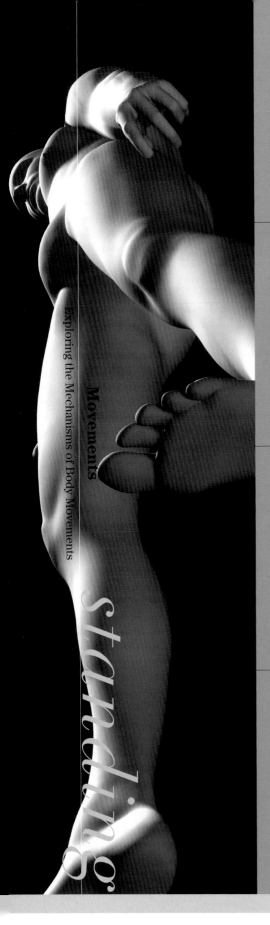

This is not a timed sequence, but a variety of different ways to stand. We'll investigate the special characteristics of these stances and discuss how impressions change a lot depending on the gender of the person doing the standing.

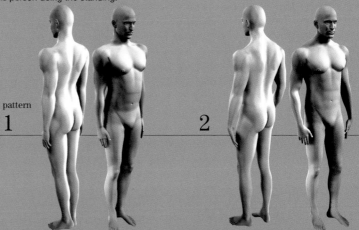

pattern

1

2

standing

the still action

Now let's turn to an action that doesn't have any clear movements connected with it—standing. Perhaps standing shouldn't be thought of as an action, but rather as a mode of expression using stances and postures. But there's no question that many body parts work to maintain balance when we stand. Let's take a long look at some different angles of people simply standing around.

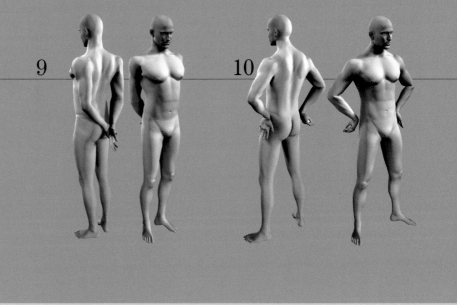

9

10

S

standing

Movements

Exploring the
Mechanisms of
Body Movements

Part 2

standing

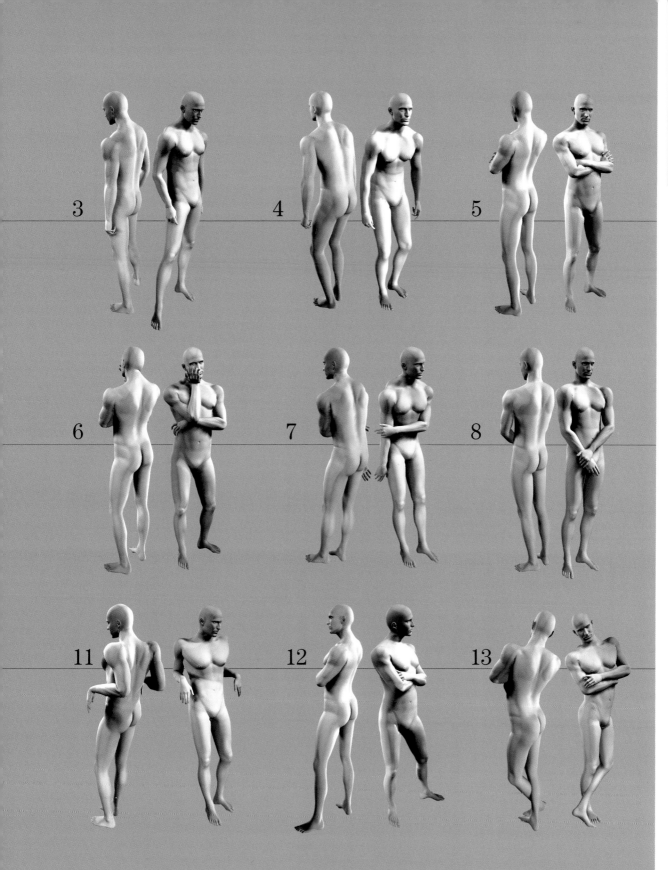

3

4

5

6

7

8

11

12

13

1

act of standing upright

Here the person is standing at attention. This is not a routine stance. The heels are aligned in the middle, and the feet angle slightly to the left and right. The knees are straight, but they are not flat. The hips are slightly forward. The upper part of the stomach, where the ribs begin, is jutting out. The back is arched, and the chest is facing up. The chest is tense, and both arms are stretched out along the sides of the body.

The palms of the hands are slightly below and at an angle from the curve of the buttocks. The chin is drawn in and pointing slightly down. The mouth naturally stays closed. This posture uses a lot of muscles.

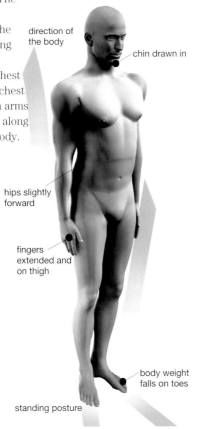

direction of the body

chin drawn in

hips slightly forward

fingers extended and on thigh

body weight falls on toes

standing posture

2

actual standing posture and balance control ☛1

This is a more relaxed stance than in frame 1. The basic rule for controlling the posture—finding the center of gravity—is important to remember. We should be aware of the relation among three points: where the body connects with the ground, the center of gravity just below the navel, and the central

point for balance at the base of the head. If balance is destroyed in these spots, the person will fall over, which is why we control our balance when we stand with a slightly collapsed posture. Neither the way the feet stand on the ground nor the way the shoulders fall back is symmetrical.

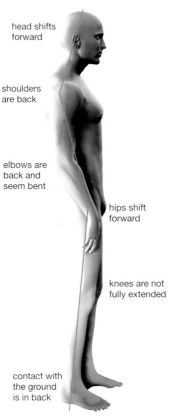

head shifts forward

shoulders are back

elbows are back and seem bent

hips shift forward

knees are not fully extended

contact with the ground is in back

the shape of balance control (side)

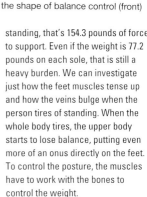

statics of the arms

points controlling the upper body

overall center

contact with ground

the shape of balance control (front)

Finding a comfortable posture to avoid fatigue

Although the act of standing keeps the body still, it is not restful like sitting or lying down. Muscles are somewhat tense as they work with the bones to support the body parts. It is the same as exercise. Standing for a long time takes both physical and mental strength. That's why a standing person will shift a lot. Imagine sitting in the *seiza* position, and it's easy to understand that when people hold any posture for a long

time, inconveniences such as pressure in the blood vessels build up, making it necessary to shift. This cycle is quicker when standing than when sitting. Remember that just standing alone can tire a person, so we try to find a slightly more comfortable posture.

☛1. Standing—the three-point arch on the soles of the feet

A three-point arch is formed on the soles of the feet with points at the base of the big toe and the base of

the small toe meeting the point at the heel through the arch. This three-point structure also allows us to stand for a long time. Think of the heel as the main axis and the toes as the springs.

☛1. Standing posture—the soles of the feet and muscles

No matter what sort of standing posture a person has, it is important not to forget that the soles of the feet are supporting the whole body. If a man who weighs 154.3 pounds is

standing, that's 154.3 pounds of force to support. Even if the weight is 77.2 pounds on each sole, that is still a heavy burden. We can investigate just how the feet muscles tense up and how the veins bulge when the person tires of standing. When the whole body tires, the upper body starts to lose balance, putting even more of an onus directly on the feet. To control the posture, the muscles have to work with the bones to control the weight.

S
standing

Movements

Exploring the
Mechanisms of
Body Movements

Part 2

standing

3
most weight on one foot: to keep standing longer ☜2

Over a long period of time, when a person tries as much as possible to avoid muscle fatigue and rely on the bone structure, he will shift his body weight somewhere. For example, he may shift his weight to either foot. In the depiction here, the model shifts his weight to the left; he'll probably shift it to the right after some time. By sharing the burden, one leg rests as the other works, allowing people to stand for a long time.

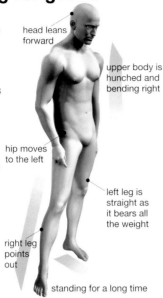

head leans forward

upper body is hunched and bending right

hip moves to the left

left leg is straight as it bears all the weight

right leg points out

standing for a long time

4
shoulders rounded: irregular strain on the muscles

We'll use the example of an extremely hunched back here. The hips are pulled back. The back is rounded, and the shoulders hang weakly. The feet are flopped down on the floor, and both heels are bearing the body weight. If the heels stop bearing the weight, the load will weigh on the joints. The hunched back makes everything about the person seem slack. The hands and feet do not seem aware, and they move in a disorderly fashion. Nothing seems symmetrical in this sort of stance.

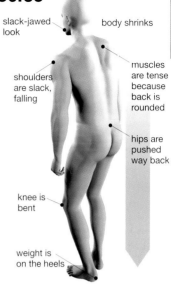

slack-jawed look

body shrinks

shoulders are slack, falling

muscles are tense because back is rounded

hips are pushed way back

knee is bent

weight is on the heels

5
arms folded: sex differences in the posture ☜3

This is the standing posture with the arms folded. It is said that when the arms are crossed, the muscles in the arms and shoulders become tense and stimulate brain activity. Men stand with their backs bent to the rear and rest their arms on the torso. Perhaps because women have breasts, they often lightly cross their arms at a lower point.

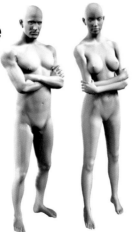

high position, firm crossing of arms

low position for crossing arms

6
chin propped on one hand: statics of the posture

The upper arm serves as a cane as the arms prop up the weight of the head. The chin rests on the palm of the right hand at the base of the thumb. The right elbow is on top of the left wrist. The left arm is across the top of the stomach. Take a close look at the lines of the body at this point: the hips up through the spine are bent three-dimensionally. The upper body, which is thrust out, leans back and to the left above the dividing line where the elbow and wrist meet.

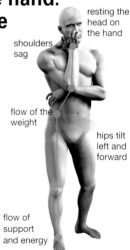

resting the head on the hand

shoulders sag

flow of the weight

hips tilt left and forward

flow of support and energy

☜1. Trouble-free standing posture

The feet are comfortably spread to either side, and the knees are not stretched out completely. Basically, the hips are pushed forward, while the upper half of the body leans back. The head is also pushed forward. From the side, the body almost forms an S. The arms are hanging at the sides, but the elbows are bent a little—the weight of the forearms pushes the arms back a bit. There's

no energy in the hands, but the joints in the fingers are bent a little, creating a gently curving line. The parts of the muscles that do the actual work are on both sides of the fingers, and when the hands aren't exerting any energy, the muscles return to the middle of their range.

☜2. Standing for a long time— the shape of different parts

When the hips lean to the left, the upper half of the body tries to keep

the shoulders level as it bends. But the left arm is straighter than the right because the latter bumps into the leg. The straight left arm pulls the shoulder down a little. The upper body slopes gently in the back. The shoulders narrow a bit in the front, and the back is a tad hunched. The neck is stretched pretty far forward, and the head is cast down.

☜3. Breaking up the weight of the arms by crossing them

The main reason for crossing the arms when standing is that they are simply in the way. By crossing them, the weight of the arms is lifted, set on the torso, and shared by the shoulders. The issue of whether the right or left arm goes on top is not strictly decided by gender—it's just a matter of how the brain works. It's something left to individual preference and awareness.

7

leaning back with shoulders against a wall

This is the posture with the back leaning against a wall. The legs are stretched tight, but not that many muscles are working. The lower the hips are against the wall the more energy is needed in the hips and at the points touching the wall to keep the body up, and the sharper the angle of the legs; the higher the hips, the more the weight is supported by the legs.

hunched back

getting the arm out of the way

The back of the hips leans against the wall.

Both feet are forward as the body leans against the wall.

8

wrist of one hand in the other in front of the body

This is a posture seen a lot in the hotel and restaurant business. The arms are crossed in the front. The chin is drawn in. It looks like a humble posture and maybe even an attractive one. A posture such as this one gives the other people present a specific impression. This is not a stance struck purely for functional reasons. It's a stance that was created within the culture to adhere to rules of etiquette.

The chin is drawn in.

The left arm attaches to the right wrist.

The hips are slightly forward.

The legs are slightly apart.

9

wrist of one hand in the other behind the body

Once a person puts the arms behind his back, a complete change can come over him, making him seem overbearing. This posture generally does not go over well in the hotel, restaurant, or entertainment business. It's a posture seen in calm adults. Put a little extremely, it's a stance that robs the body of its freedom. It requires control, so it's not likely to be seen in a child in the fever of youth. In fact, it's not seen much in general. The arms can't operate well from behind the back, which means they may not be able to respond if a quick transformation takes place.

The chin is drawn in; the head looks down.

The shoulders are drawn deeply back.

The upper body leans back.

The top of the stomach juts out.

The hips push forward.

The feet are drawn back.

10

hands on the waist: sex differences in the posture

The hips are pushed forward and the hands are on the hips in this stance. The poses differ a little from men to women. If the thumbs are on the back, the chest naturally fills out. The legs are far apart. This posture gives off an image of pride. If the forearms are twisted around in the opposite way, the shoulders hunch a little and a tired image is created. The twisting of arms and legs differs from male to female. With many different postures, twisting a limb a certain way can establish the gender.

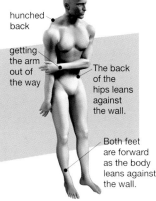

Men: The hand may make an ordinary grip as the wrist touches the hip.

Women: The hand may make a reverse grip as the wrist touches the hip.

11

leaning back with elbows on a railing

This stance is only seen if a convenient support is behind the person. Both arms support a lot of the upper body's weight, decreasing the onus on the legs. Watch for signs where a person puts more weight on the right arm, as in the diagram, or the left. This changes the shape of the body. The elbows are back, pointing toward the center. The upper arms point out. This is because the elbows want to be close to the power center in the shoulders.

The head is forward.

Both shoulders are raised.

The elbows support the upper body.

The elbows head in different directions.

The hips rise to the left.

The knee moves back and bends.

Both feet are forward.

12

arms folded and weight shifted: compound posture

The poses introduced so far are a combination of many postures. In this one, the arms are crossed, and the weight is shifted to one leg. Note the twist in the hips. The hips are level when the weight is equally shared by both legs. When someone stands for a long time, he will usually end up favoring one leg or the other.

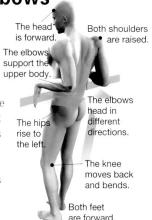

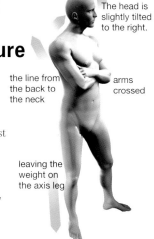

The head is slightly tilted to the right.

the line from the back to the neck

arms crossed

leaving the weight on the axis leg

13

leaning with one shoulder against a wall

The shoulder rises as it leans against the wall.

direction of the upper body

The hips lean toward the wall.

the energy flow

Only the toes touch the ground.

This posture puts the whole weight of the body against the wall. The right leg forms the axis, and the left leg curls around it. The right leg is firmly on the ground; the toes on the left leg are not so firmly planted. The hips are twisted a lot and facing the wall. The feet are also naturally facing the wall. The spine is bending quite a bit toward the right, and the left shoulder is sticking out. The area from the left shoulder to the elbow is against the wall; the whole body is deposited here. The neck falls on the right side of the body's median line. Pay special attention to the right foot, which is in contact with the ground; the hip directly above; and the left shoulder, which is leaning on the wall.

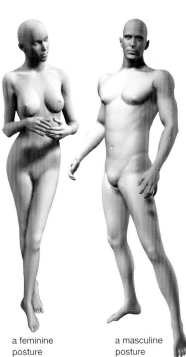

a feminine posture

a masculine posture

ETCETERA
Gender differences can come to life in postures

Standing, the subject of this chapter, tends to result in men taking frank, open poses that exude strength, while women favor inward, delicate stances. This is all coming from our ideas of ourselves and our genders, as well as from societal ideas about gender and reflexive actions. If we break down the differences further, we can see congenital differences, such as the way the joints in women can bend so far backward. Also, once you have mulled over gender differences, you have to prepare for the even bigger task of investigating individual differences.

The various transformations of the action called "standing"

In the course of human progress, standing on two feet has given us freedom and a complex array of movements. Biologically speaking, we've lost a lot, too. We have to toil quite a bit to control our bodies; in fact, we have a more difficult job than our four-legged friends. To realize the complex collaboration of many different parts, humans have highly developed motor skills.

To grasp the complex movements of the human body, we must of course investigate the reflexive, instinctive and biological actions, but also the feelings and culture of a person. This book thinks the "action of habits" depicts personality. It is difficult to decide whether this is a biological or congenital matter, or whether it is cultivated from a cultural background or just something in one's personality. But it is certainly one of the keys to depicting movement.

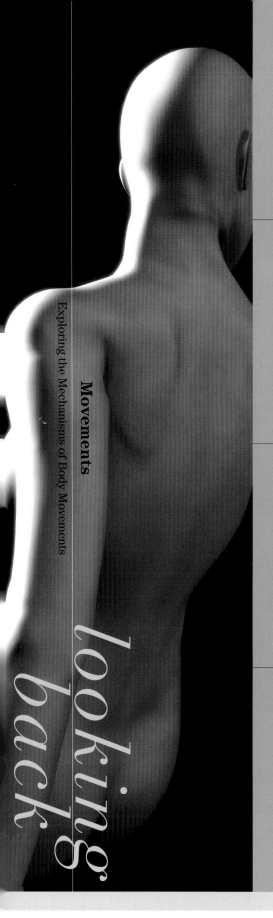

looking back

looking back

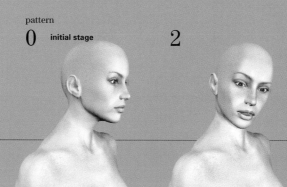

pattern

0 **initial stage**   2

looking back

movement and awareness of the eyes

Generally speaking, there are three different ways to look back: an instinctive way that originates in the desire for self-preservation; an intentional way when someone tries to look at something; and a reflexive way when someone calls our name, for example. We'll focus here on the reflexive way of looking back.

3   4   5

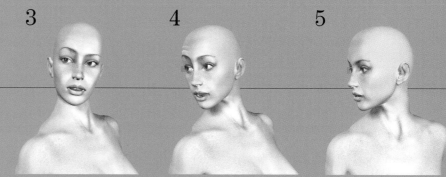

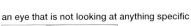

l
looking back

Movements

Exploring the
Mechanisms of
Body Movements

Part 2

looking
back

0

in the beginning

Let's assume that while a person is sitting down, someone calls out to her, and she turns to see who it is. The action is a short two seconds or so. We won't be portraying this action in frames; instead, we will simply focus on the significant movements as they occur over time.

basic posture at
the beginning

1

before the action

This is before the person looks back. The back is slightly sloped and the shoulders sag—it's a normal position. The eyes aren't really looking at anything specific; they look out into space, often straight ahead and down at an angle. Actually, it is rare for people to look at nothing; the eyes are usually restlessly following something. That doesn't mean they are always aware of what they are looking at, but the head and eyes are constantly looking around.

an eye that is not looking at anything specific

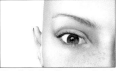

The first direction usually taken by an eye that isn't looking at anything is a spot diagonally below the center.

The eyes are still not looking at anything.

The neck leans forward.

2

initiating the action ☛1

The head begins to rotate in the direction of the call. Usually, the jaw leads the way. The head tilts a little in the opposite direction—in this case, to the left. In the depiction here, we've assumed that the woman was suddenly called. Her mouth was closed (not in a conscious way), but as she turns, it opens slightly. At this stage, the mouth isn't always open, but for an instant, as the head rotates, it seems a natural pose.

The chin is drawn in a bit. The eyes are focused on nothing in particular, but the line of vision is diagonally down and to the right. The woman has no special expression. She has just begun to turn her head.

line of vision in the middle of moving

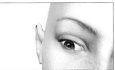

As the eyes turn back to see who is calling, the line of vision drops diagonally down and to the right.

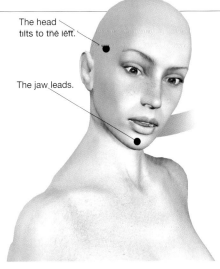

The head tilts to the left.

The jaw leads.

☛**1. Looking back**

Instinctive actions are very quick and difficult to catch. For example, if a loud noise occurs, we infer that danger is near and turn as quickly as possible in the direction of the sound. To that point, it is still a subconscious activity. But after that, the person considers the situation, estimates the danger, and plans the next action. When we engage in a subjective action like looking back for something, we don't need to rush as much as we do with an instinctive action. If there is nothing special about the circumstances, we can take some time to make the move.

The subject has not yet been identified as the person begins to look back, so her eyes drift in space for a while. When a person reflexively looks back, the subject has made its general position clear by making noise; the eyes estimate the source of the noise and naturally catch up to it.

However, when looking back, we sometimes glance at another spot, or fully turn to look squarely at the object of our attention, etc. We can also look back using only our neck and head, but the weight of the neck alone will end up shifting the body weight. When we consider the aftermath, or if we just want to capture the action precisely, it is preferable to have the upper body move as well. The spine is connected to the bones in the neck, which is why the shoulders also move.

3
straightening the upper body

The upper body straightens. The head was tilted slightly to the left, but it rises as the whole body moves back. The biggest change occurs in the eyes: they were gazing down, but suddenly they look up and to the right. They haven't found the subject yet, so they aren't yet focused on anything. They catch the subject in the center of their vision, so the eyes tend to cross. When they tilt up, the eyebrows naturally follow. As the person consciously tries to see behind her, the eyebrow closest to the subject—in this case, the right side—rises. As the chin turns sideways, the skin underneath stretches, naturally opening the mouth. But the mouth doesn't always open at this point.

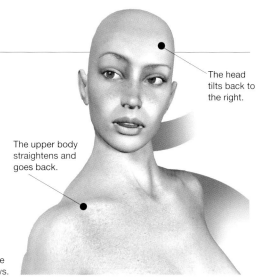
The head tilts back to the right.

The upper body straightens and goes back.

eye looks for subject

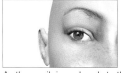

As the pupil rises sharply to the upper right, so do the eyebrows.

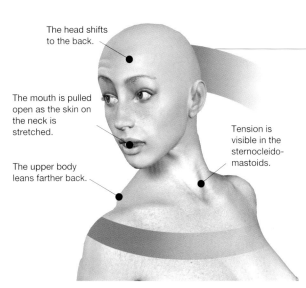
The head shifts to the back.

The mouth is pulled open as the skin on the neck is stretched.

The upper body leans farther back.

Tension is visible in the sternocleido-mastoids.

4
catching the target with the eyes

The upper body leans farther back. The neck stretches to nearly its limit. This is the moment with the biggest changes. The skin on the neck is strained, and wrinkles appear. It's an unnatural posture, with the sternocleidomastoids in the neck straining under the burden of supporting the head. The eyebrows also rise, and creases appear on the forehead. The eyes open wide as the subject is seen. The head leans farther back, and the mouth opens wide.

catching the target with the eyes

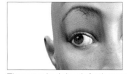

The eyes look back farther and open wide; the eyebrows rise, and creases appear on the forehead.

5
taking a more comfortable posture

To this point, the person hasn't been thinking about her movements. She's been taken along by the course of events until her upper body is in an unnatural posture. The upper body is firmly straightened after leaning back too much. The body turns at the waist to face the subject. The spine is slightly bent to the left, and the right shoulder is higher than the left. The neck is still turned back. The chin draws in as the upper body straightens. The subject is recognized, and the upper body is once again under control. The mouth shuts. The line of vision has also been raised, so the eyes, which were looking up, are in a comfortable position. The woman can see the person who called out to her, and she starts to be conscious of being seen. The independent action of looking at the subject is combined with a subconsciously defensive look that comes from being watched.

the eye when being seen

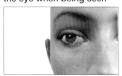

Once a person returns to a posture that makes it easier to see, the eyes view the subject, and at the same time, react to being seen.

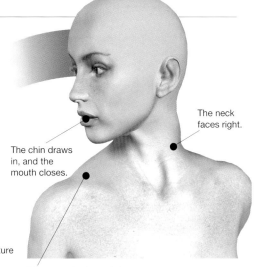
The neck faces right.

The chin draws in, and the mouth closes.

The upper body straightens by moving forward.

Afterword

My goal in making this book was to create a how-to manual that would present the natural movements of the human body, with all of the basic anatomical knowledge and applications necessary for designers. When I began writing, I wanted to find an answer to the rather vague question, What is animation? I also wanted to continue learning about the movements of people.

Early in my career, when I had just started to copy the human form as a computer graphics designer, I tried to make my drawings appear as if they were alive, but I didn't even consider the differences between animal and human movements. As time passed, I began to ask some important questions: What are bones like? How do muscles function? How are individual body parts used? I then saw that there are many differences in the way we as humans perceive and move through the world around us. I knew, that in order to be successful, I would need a firm grasp of the human anatomy.

It is said that to know a person, one should pay attention to his or her speech and conduct. Words are key elements to understanding someone, but often the eyes can say more than the mouth. By watching someone's gestures, you can distinguish one person from another; and if your relationship deepens, you can even read their mind. Once we discover a character's thoughts and desires, and the way these thoughts emerge in their daily habits, we can begin to create more natural physical expressions.

Working on this book has given me the chance to answer all of those questions I asked as an ambitious designer, and I hope that it will provide you with the same important guidelines on your long journey to discovering the vast universe of the human body and its movements.

Takashi Iijima

Index

Index